Timm Tanzeglock

A Novel Taylor-Couette Bioreactor for Mammalian Cell Cultivation

Timm Tanzeglock

A Novel Taylor-Couette Bioreactor for Mammalian Cell Cultivation

The cellular response to environmental stress and its use for the design of a bioreactor for the cultivation of shear sensitive cells and tissues

Südwestdeutscher Verlag für Hochschulschriften

Impressum/Imprint (nur für Deutschland/ only for Germany)
Bibliografische Information der Deutschen Nationalbibliothek: Die Deutsche Nationalbibliothek verzeichnet diese Publikation in der Deutschen Nationalbibliografie; detaillierte bibliografische Daten sind im Internet über http://dnb.d-nb.de abrufbar.
Alle in diesem Buch genannten Marken und Produktnamen unterliegen warenzeichen-, marken- oder patentrechtlichem Schutz bzw. sind Warenzeichen oder eingetragene Warenzeichen der jeweiligen Inhaber. Die Wiedergabe von Marken, Produktnamen, Gebrauchsnamen, Handelsnamen, Warenbezeichnungen u.s.w. in diesem Werk berechtigt auch ohne besondere Kennzeichnung nicht zu der Annahme, dass solche Namen im Sinne der Warenzeichen- und Markenschutzgesetzgebung als frei zu betrachten wären und daher von jedermann benutzt werden dürften.

Verlag: Südwestdeutscher Verlag für Hochschulschriften Aktiengesellschaft & Co. KG
Dudweiler Landstr. 99, 66123 Saarbrücken, Deutschland
Telefon +49 681 37 20 271-1, Telefax +49 681 37 20 271-0, Email: info@svh-verlag.de
Zugl.: Zurich, ETH Swiss Federal Institute of Technology Zurich, Diss., 2008

Herstellung in Deutschland:
Schaltungsdienst Lange o.H.G., Berlin
Books on Demand GmbH, Norderstedt
Reha GmbH, Saarbrücken
Amazon Distribution GmbH, Leipzig
ISBN: 978-3-8381-0705-9

Imprint (only for USA, GB)
Bibliographic information published by the Deutsche Nationalbibliothek: The Deutsche Nationalbibliothek lists this publication in the Deutsche Nationalbibliografie; detailed bibliographic data are available in the Internet at http://dnb.d-nb.de.
Any brand names and product names mentioned in this book are subject to trademark, brand or patent protection and are trademarks or registered trademarks of their respective holders. The use of brand names, product names, common names, trade names, product descriptions etc. even without a particular marking in this works is in no way to be construed to mean that such names may be regarded as unrestricted in respect of trademark and brand protection legislation and could thus be used by anyone.

Publisher:
Südwestdeutscher Verlag für Hochschulschriften Aktiengesellschaft & Co. KG
Dudweiler Landstr. 99, 66123 Saarbrücken, Germany
Phone +49 681 37 20 271-1, Fax +49 681 37 20 271-0, Email: info@svh-verlag.de

Copyright © 2009 by the author and Südwestdeutscher Verlag für Hochschulschriften Aktiengesellschaft & Co. KG and licensors
All rights reserved. Saarbrücken 2009

Printed in the U.S.A.
Printed in the U.K. by (see last page)
ISBN: 978-3-8381-0705-9

Contents

Contents		**1**
Abstract		**3**
Zusammenfassung		**5**
1	**Introduction**	**7**
1.1	Outline 9	
2	**Materials and Methods**	**10**
2.1	Cell Culture 10	
	2.1.1 Cell Lines 10	
	2.1.2 Culture Media 10	
	2.1.3 Cryopreservation and Thawing of Cells 11	
2.2	Analytic Techniques 11	
	2.2.1 Cell Count, Size, Viability and Microscopy 11	
	2.2.2 Metabolism Monitoring via HPLC Analysis 12	
	2.2.3 Colorimetric SEAP Assay 14	
	2.2.4 DNA Measurement 14	
	2.2.5 Detection of Apoptotic and Necrotic Cell Death 15	
	2.2.6 Bradford Protein Assay 19	
2.3	Visualization of the Flow Field 20	
2.4	$k_L a$ Measurement 20	
2.5	Mixing Time 21	
2.6	Experimental Procedure for Hydrodynamic Stress Exposure 22	
3	**Response of Mammalian Cells to Environmental Stress**	**25**
3.1	Introduction 25	
	3.1.1 Cell Damage due to Shear Exposure 25	
	3.1.2 Flow Devices and Fluid Flow Characterization 30	

3.2 Results and Discussion 34
    3.2.1 Cell Response to Laminar Simple Shear Flow..34
    3.2.2 Extensional Flow - Flow Field Characterization...40
    3.2.3 Cell Response to Extensional Flow (Short Term)...47
    3.2.4 Cell Response to Extensional Flow (Long Term)...56
3.3 Conclusions 58

4 Development of a Novel Taylor-Couette Bioreactor   **60**

4.1 Introduction 60
    4.1.1 Bioreactor Design Considerations..60
    4.1.2 State of the Art in Biopharmaceutical Development and Production 62
    4.1.3 Design and Construction of the LTC Reactor 65
    4.1.4 Design of the LTC Bioreactor 71
    4.1.5 Stirred Tank Reactor..73
    4.1.6 Numerical Details of the CFD Simulations..75
4.2 Results and Discussion 76
    4.2.1 Hydrodynamics and Flow Field Characterization..76
    4.2.2 Mixing Time 79
    4.2.3 Mass transfer characteristics...80
    4.2.4 Membrane Lifetime and Reusability 82
    4.2.5 Cell Cultivation...83
4.3 Conclusions 91

5 Conclusions   **92**

6 Outlook   **94**

7 Bibliography   **96**

Abstract

In the pharmaceutical industry mammalian cell lines gained more and more attention in the last decade and are nowadays the primary expression system in the production of therapeutic recombinant proteins. Despite its diverse drawbacks, the stirred tank reactor (STR) is usually the system of choice and serves as major workhorse for most applications. To overcome the limitations faced in established processes, improved bioreactors are needed to provide a more favorable *in vitro* environment to mimic *in vivo* conditions.

In the first part of this work the influence of shear flow on the cell physiology is investigated to identify and anticipate the impact of this process parameter during cell cultivation. The major goal is to determine whether the type of shear flow, to which cells are exposed, influences the initiation of cell death. Two flow devices are employed to impose distinct hydrodynamic flow fields: a uniform steady simple shear flow and an oscillating extensional flow. The cellular response is evaluated by measuring the release of intracellular components into the culture supernatant, as indicator for membrane rupture and necrotic death, and by means of fluorescence activated cell sorting, to distinguish between necrotic and apoptotic cell death. It is shown that mammalian cells, indeed, distinguish between discrete types of flow and respond differently. They will enter the apoptotic pathway when subjected to low levels of hydrodynamic stress in oscillating, extensional flow. In contrast, necrotic death prevails when the cells are exposed to low levels of hydrodynamic stresses in simple shear flow or to high levels in extensional flow. However, under prolonged exposure to extensional flow cells suffer severe necrosis even under shear stresses comparable to the threshold obtained using simple shear flow. The determined threshold values at which cells will enter the respective death pathway should be avoided when culturing cells to enhance culture longevity and productivity.

Based on these studies, in the second part of this work, a novel bioreactor for the cultivation of shear sensitive cells and tissues is introduced. The proposed device is based upon a lobed Taylor-Couette reactor, which was developed recently in our laboratory. The unit consists of two concentric cylinders of different diameter. The inner one is polygonal-shaped and furnished with a silicone membrane for oxygenation. The cells are cultivated within the annular gap, where the liquid is set in motion by the rotation of the inner cylinder. It is shown that the proposed device is featuring good mass transfer characteristics at low shear values both in maximal magnitude as well as distribution. Further, the lobed Taylor-Couette bioreactor emerged superior to a common stirred tank reactor regarding cell cultivation: It is proven that the lag phase after inoculation is shortened, the maximum cell density as well as the growth rate is increased, and that the volumetric productivity is enhanced. Moreover, the lobed Taylor-Couette reactor introduces ten times more energy to the system, and therefore enhancing mass transfer with respect to the STR, without sacrificing cell viability or productivity. Additionally, cell cultivation without the medium supplementation with cell protective or antifoam reagents is feasible. Especially nowadays, where product integrity and consistency are of uttermost importance and strongly controlled by the regulatory authorities, the lobed Taylor-Couette reactor provides a suitable tool for biopharmaceutical research, development, and production. Possible applications could be found in the cultivation of cell suspensions, cell aggregates, large plant cells, adherent cells on microcarriers, shear-sensitive biocatalysts, stem cells or tissue scaffolds.

Zusammenfassung

Die Kultivierung von Säugetierzellen zur Produktion von pharmazeutischen Proteinen erlangte in den letzten Jahrzehnten große Aufmerksamkeit. Während der Prozessentwicklung und in der industriellen Produktion werden hierbei größtenteils Rührkessel verwendet. Nichtsdestotrotz unterliegt dieses System verschiedenen Limitationen, so dass die Entwicklung von alternativen Bioreaktoren im Fokus wissenschaftlicher Arbeit steht.

Im ersten Teil dieser Arbeit wird eine detaillierte Studie über das zelluläre Verhalten bei Stressbeanspruchung im Kulturmedium durchgeführt. Mittels zweier hydrodynamischer Modelsysteme werden Säugetierzelllinien gezielt definierten Scherbeanspruchungen ausgesetzt: einerseits einer gleichförmigen, einfachen Scherströmung in einem Rotationsviskometer und anderseits einer oszillierenden Dehnströmung, generiert in einer sich verengenden Kapillare. Hierbei wird aufgezeigt, dass bei Überschreiten einer kritischen Scherspannung die Zellen makroskopisch beschädigt werden und Nekrose eintritt. In Abhängigkeit von der auftretenden Strömungsform, unterliegen die Zellen unterschiedlichen Mechanismen des Zelltodes. Setzt man die suspendierten Zellen wiederholt einer schwachen Dehnströmung aus, tritt Apoptose (programmierter Zelltod) ein. Im Gegensatz dazu dominiert im laminaren Strömungsprofil eines Rheometers Nekrose. Anhand der ermittelten kritischen Grenzwerte können Rückschlüsse auf die Scherempfindlichkeit der Zelllinien gezogen werden und diese bei der Entwicklung und Optimierung von biopharmazeutischen Produktionsverfahren genutzt werden.

Im zweiten Teil dieser Arbeit werden die gewonnenen Erkenntnisse zur Scherempfindlichkeit der Zelllinien verwendet, um einen neuartigen Bioreaktor zu entwickeln und in Betrieb zu nehmen. Konzeptionell basiert der Reaktor auf einem Taylor-Couette System, welches am Institut für Chemie- und Bioingenieurwissenschaften der ETH Zürich entwickelt wurde. Der Reaktor besteht aus zwei konzentrischen Zylindern unterschiedlichen Durchmessers, wobei der Innere einen polygonalen Querschnitt aufweist. Im Ringspalt wird die Zellkultivierung durchgeführt, wobei die Durchmischung des Mediums durch Rotation des Innenzylinders gewährleistet wird. Zur Sauerstoffversorgung der Zellsuspension wird eine Silikonmembran verwendet, die auf dem perforierten Innenzylinder befestigt ist, welcher von Innen unter Druck gesetzt wird. Hydrodynamisch betrachtet zeichnet sich der beschriebene Reaktor vor allem durch eine enge Verteilung der Energiedissipation im Reaktionsvolumen und einer verbesserten Mischzeit aus. Nach erfolgter Charakterisierung des Reaktors bezüglich Mischzeit, Massentransfer und Energiedissipation wird seine Tauglichkeit zur Zellkultivierung nachgewiesen. Als Vergleichsreaktor wird hierbei ein kommerzieller Rührkessel verwendet. In Bezug auf Wachstumsrate, maximal erreichbarer Zelldichte und Produktionsrate erweist sich der Taylor-Couette Reaktor dem Rührkessel überlegen. Außerdem erlaubt das vorgestellte System einen höheren Energieeintrag bei gleichzeitig geringerer Scherbeanspruchung und den Betrieb ohne medienergänzende Zellschutzreagenzien, und stellt somit eine vielversprechende Alternative zum konventionellen Rührkessel dar. Mögliche Einsatzbereiche reichen von der Kultivierung von Zellen in Suspension, als Aggregate oder auf Microcarriern bis hin zu scherempfindlichen Biokatalysatoren, Stammzellen oder Gewebekulturen.

Chapter 1
Introduction

In the pharmaceutical industry various host cell systems for the production of therapeutic recombinant proteins are used. Due to their ability to achieve appropriate posttranslational modifications and proper protein folding, mammalian cell lines gained in the last decade more and more attention and are nowadays the primary expression system. The number of approved pharmaceuticals produced in mammalian cell lines has increased in the past years tremendously and makes now around 55 % of all products on the market (Walsh, 2006). Here, especially Chinese Hamster Ovaries (CHO), Baby Hamster Kidney (BHK), murine myeloma (NS0), and human embryonic kidney (HEK) cells are dominating (Wurm, 2004). The recombinant protein therapeutics market value showed double digit growth during the last years and increased from 14.5 billion US$ (1998) (Lawrence, 2005) to 35 billion US$ (2004) and is forecasted to exceed 70 billion US$ in 2010 (Pavlou and Reichert, 2004; Walsh, 2006). In this environment of a fast growing and competitive market, governmental price control and rising quality requirements made by the authorities, further research to ensure sufficient production capacity and constantly improved product quality is needed (Dickson and Gagnon, 2004).

The expansion of all types of cells requires suitable bioreactors to provide an optimal *in vitro* environment, mimicking *in vivo* conditions. Ideally, the bioreactor should provide and maintain the highest degree of environmental uniformity, avoiding concentration gradients of nutrients, toxic metabolic byproducts and oxygen, without featuring critical levels of shear stress (Nienow, 2006). Next to the necessity of providing a gentle environment for the cultivation of shear sensitive cells and tissues, there are three other main concerns faced during pharmaceutical production, namely increasing the maximal cell density and viability, maximizing the specific productivity, and enhancing the longevity and stability of the culture (Butler, 2005; Wurm, 2004).

Over the years researcher have investigated several methods to reduce cell damage due to hydrodynamic forces and proposed bubble free oxygenation methods (Ducommun et al., 2000; Lehmann et al., 1987; Qi et al., 2003; Schneider et al., 1995), low-shear impellers and novel agitation systems (Cervantes et al., 2006; Nienow et al., 1996), and the utilization of cell protective agents like Pluronic F-68 and fetal calf serum (Chattopadhyay et al., 1995; Keane et al., 2003; Kunas and Papoutsakis, 1989; Michaels and Papoutsakis, 1991; Murhammer and Goochee, 1988; Wu, 1999). The latter becomes ineligible because of regulations made by the authorities and process considerations from the pharmaceutical industry (prion contamination, lot-to-lot variations, additional contaminants during downstream processing, difficult licensing procedures, etc. (Even et al., 2006)). Moreover, the application of cell protective agents such as Pluronic F-68 exhibit also major drawbacks connected to a decreased mass transfer (Moreira et al., 1995; Murhammer and Pfalzgraf, 1992), and additional problems in the downstream purification process (Heath and Kiss, 2007) like a decreased protein binding capacity of chromatographic resins (Low et al., 2007). Another point of interest emerges when looking at the downstream-part of the bio-pharmaceutical production scheme, where intracellular components released by damaged cells, will contaminate the culture medium and complicate the purification process. Therefore, the upstream process has to be designed carefully in order to expose the cultured cells to a minimum level of debilitative forces.

Despite all efforts, the existing cultivation units, like stirred tank reactors or rotating wall vessels, suffer from diverse drawbacks (Al-Rubeai et al., 1995; Alexopoulos et al., 2002; Arratia et al., 2004; Chisti, 2001; Garciabriones and Chalmers, 1994) and make the development of improved bioreactors essential. To establish a suitable culture system, the cell's physiology and responsiveness to environmental stress has to be evaluated to identify and anticipate the impact of this process parameter during cell cultivation.

Chapter 1 - Introduction

1.1 Outline

This work pursues the following objectives:

- **Investigation of the Cellular Response to Environmental Stress**

  The main goal of the first part of this study is to investigate whether the type of shear flow, to which cells are exposed, influences the initiation of cell death. Moreover, critical values of the hydrodynamic stress leading to necrosis using industrially relevant cell lines growing in culture medium without additional supplements shall be determined. Since other mechanisms, in particular apoptosis could be present and causing cell death, the focus will be on identifying conditions for which this death mechanism dominates. Two flow devices are employed to impose accurate hydrodynamic flow fields: a uniform steady simple shear flow generated in a cup and bob rheometer and an oscillating extensional flow created in a sudden contraction at the entrance to a capillary. The obtained threshold values at which cells will enter the respective death pathway should be avoided when culturing cells for recombinant protein production.

- **Development of a Bioreactor for the Cultivation of Shear Sensitive Cells and Tissues**

  Considering the findings of the studies made before, a novel culture system, based on a Taylor-Couette device, is designed and constructed in order to establish a superior cultivation environment for shear sensitive cell lines and tissues enhancing culture longevity, cell density and volumetric productivity. Emphasis is given to the problems connected to bubble aeration and the heterogeneity of the microenvironment of commonly used systems, like stirred tank reactors. The qualification of the bioreactor for cell culturing is demonstrated by the cultivation of protein-free growing Chinese Hamster Ovaries and the process performance is compared to that of a commercial bench-top stirred bioreactor. Furthermore, the new production process is designed in order to enable cell cultivation without the addition of cell protective agents and give the flexibility of batch and continuous operation mode.

Chapter 2
Materials and Methods

2.1 Cell Culture

2.1.1 Cell Lines

Since Chinese Hamster Ovaries (CHO) and Human Embryonic Kidney (HEK) cells are most frequently used in recombinant protein production (Butler, 2005; Wurm, 2004) they are chosen as representative model systems for all experiments. The CHO-K1 descendent CHO easyC (Cell Culture Technologies, Switzerland) is transfected to constitutively express secreted alkaline phosphatase (SEAP) (Schlatter et al., 2002), while the HEK ebna 293 cells produce the Epstein-Barr virus (EBV) nuclear protein (Geisse and Kocher, 1999). Both cell lines are kindly provided by Prof. Fussenegger (ETH Zurich, Switzerland).

The cell lines are anchorage-independently grown in T-flasks and roller-bottles at 37° C in a humidified atmosphere containing 5 % carbon dioxide. 24 hours before each experiment, the cells are split and provided with fresh medium. Immediately before the experimental start, cells are pelleted and resuspended in fresh medium according to the appropriate final cell concentration.

2.1.2 Culture Media

CHOMaster HTS and HEKTOR G medium (Cell Culture Technologies, Switzerland), both protein-free and without any additional supplements, are used for growing the CHO and HEK cell line, respectively. The media is provided as concentrate and prepared following the instructions of the manufacturer. Pluronic F-68 (Sigma-Aldrich, Switzerland) is used at a concentration of 0.1 % (w/v).

2.1.3 Cryopreservation and Thawing of Cells

Viable cells at a concentration of 5×10^5 cells mL^{-1} are harvested, centrifuged for 3 minutes at 1200 rpm at room temperature (RT), and resuspended in 45 % fresh media, 45 % conditioned media, and 10 % (v/v) dimethyl sulphoxide (Sigma-Aldrich, Switzerland). The cell solution is aliquoted into sterile cyrotubes and stored at -80° C overnight in a freezing container filled with isopropanol to maintain a cooling rate of -1° C per minute. Afterwards the vials are transferred for long-term storage in a -140° C freezer. For thawing, the cell solution is warmed in a waterbath and resuspended in 10 mL of fresh, pre-heated media. The solution is centrifuged for 3 minutes at 1200 rpm at RT, the pellet resuspended in fresh media and incubated at 37° C.

2.2 Analytic Techniques

2.2.1 Cell Count, Size, Viability and Microscopy

Cell count and cell diameter are measured by a CASY counter (Schärfe System GmbH, Germany). Before injection, the cell sample (100 μL) is diluted with 10 mL of pre-filtered CASYton, a weak electrolyte provided by the manufacturer. The cell viability is determined by the use of an improved Neubauer hemacytometer (Omnilab AG, Switzerland) and trypan blue (Sigma-Aldrich, Switzerland) staining. The cell growth rate, μ, is calculated through the following equation:

$$\mu = \frac{\ln X_2 - \ln X_1}{t_2 - t_1} \quad (1)$$

with X_1 and X_2 being the viable cell concentrations at time t_1 and t_2, respectively. The doubling time, t_D, follows from the above equation and is expressed as:

$$t_D = \frac{\ln 2}{\mu} \quad (2)$$

Visual inspection of the cells is performed under a light microscope (Leica AG, Germany) equipped with a digital camera (Canon Inc., Japan).

Chapter 2 - Materials and Methods

2.2.2 Metabolism Monitoring via HPLC Analysis

For the cultivation of all types of cells, a well defined culture media, containing the various essential nutrients, is needed (Eagle, 1955). During the cell's catabolism the medium components are transformed into several end-products, which could inhibit cell growth (Reuveny et al., 1986), product formation (Ito and McLimans, 1981), and protein glycosylation (Andersen and Goochee, 1995). The major waste products lactate and ammonium were found to inhibit cell growth and productivity at levels of around 30 mmol and 2 mmol, respectively (Cruz et al., 2000; Luo and Yang, 2004).

Monitoring and controlling of the culture metabolism is crucial for the success and reproducibility of any bioprocess. For this purpose a HPLC (GE, USA) analysis using an Aminex 300 x 7.8 mm HPX-87H column + guard column (both from Bio-Rad Laboratories, Switzerland) is established. The stationary phase consists of a sulfonic cation-exchange material and allows the separation of carbohydrates and amino acids in the presence of an organic acid (Ross and Chapital, 1987; Weigang et al., 1989).

First the guard column is flushed with 15 mL of eluent (0.05 mol H_2SO_4) at a flow rate of 0.25 mL min^{-1}. Afterwards the Aminex column is connected, flushed with additional 20 mL of mobile phase and slowly heated up to 40° C. After reaching the operational temperature, the flow rate is ramped up to the final value of 0.5 mL min^{-1}.

Substance	t_R / min	Linear correlation	R^2	Detector
Glucose	10.94	y = 5.20E-08x-1.78E-05	1.00	RI
Lactate	15.30	y = 1.05E-05x-2.17E-04	1.00	DAD 210nm
Sodium-Pyruvate	12.99	y = 1.98E-05x+2.92E-04	1.00	DAD 210nm
Sodium-Acetate	18.34	y = 1.99E-04x-1.10E-04	1.00	DAD 210nm
L-Glutamine	23.60	y = 3.11E-05x-1.13E-03	1.00	DAD 210nm
L-Asparagine	16.17	y = 7.48E-07x+3.20E-03	1.00	RI

Table 1 Aminex HPX-87H column calibration: runtime 40 min, flow rate 0.5 mL min^{-1}, column temperature 40° C, mobile phase 0.05 mol H_2SO_4.

The frozen samples are thawed, aliquoted and analyzed by an isocratic HPLC method. Eluted components are monitored with an ultraviolet (at a wavelength of 210 nm) and a refractive index (RI) detector connected in series. By comparing the retention time, t_R, and the peak area with various standard substances (Sigma-Aldrich, Switzerland) (see *Table 1* and *Figure 1*), single components can be analyzed and their concentrations determined.

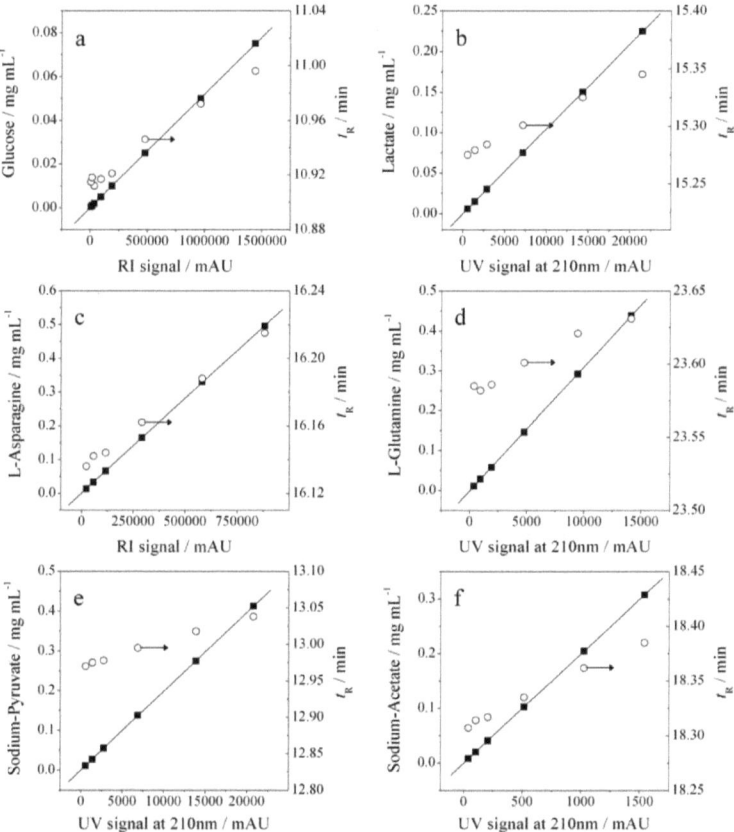

*Figure 1 Aminex HPX-87H calibration. (**a**) Glucose (**b**) Lactate (**c**) L-Asparagine (**d**) L-Glutamine (**e**) Sodium-Pyruvate (**f**) Sodium-Acetate. (○) retention time t_R (■) detector signal.*

Chapter 2 - Materials and Methods

For all detected components the calibration gives a linear signal over a broad concentration range (*Figure 1*). The change of retention time with respect to a change in concentration is negligible and the peaks are well separated from each other.

The specific substrate uptake and lactate production rate are calculated by the following equations:

$$q_{glc} = \frac{s_1 - s_2}{\left(\frac{X_1 + X_2}{2}\right)(t_2 - t_1)} \quad (3)$$

$$q_{lac} = \frac{p_2 - p_1}{\left(\frac{X_1 + X_2}{2}\right)(t_2 - t_1)} \quad (4)$$

where s_1 and s_2 are the glucose and p_1 and p_2 the lactate concentrations at time point t_1 and t_2, respectively.

2.2.3 Colorimetric SEAP Assay

SEAP activity in clarified culture supernatants is measured based on the rate of enzymatic hydrolysis of p-nitrophenylphosphate (pNPP) to p-nitrophenol. For this purpose 200 µL of culture supernatant are incubated for 10 min at 65° C to deactivate endogenous phosphatases. Occurring condensation water at the tube lid is spinned down shortly. 80 µL of supernatant are subsequently mixed with 100 µL of 2x SEAP buffer (20 mmol L-homoarginine, 1 mol diethanolamine and 1 mmol $MgCl_2$, pH 9.8). After initiating the enzymatic reaction by adding 20 µL of 120 mmol pNPP (Sigma-Aldrich, Switzerland), the light absorbance at 405 nm is monitored for 30 min in 30 s time intervals using a microwell plate reader (Tecan Trading AG, Switzerland). All assays are performed in duplicates. From the resulting slope the enzymatic activity in units per liter is calculated (Schlatter et al., 2002) and converted to µg L^{-1} by dividing by a factor of two (Ezra et al., 1983). The volumetric productivity is calculated at the time of the maximum viable cell concentration, by dividing the corresponding SEAP concentration by the culture time.

2.2.4 DNA Measurement

The release of DNA into the culture medium upon shear serves as good indicator for cell damage and membrane rupture (Croughan and Wang, 1989; Dunlop et al., 1994; Lutkemeyer et al., 1999; Racher et al., 1990; Senger and Karim, 2003). DNA quantification is done using the fluorescence-based Quant-iT DNA assay kit (Molecular Probes, Switzerland). The assay is highly selective for dsDNA over RNA and the fluorescence signal (excitation at 492 nm, emission at 535 nm) is linear in the range of 0.2-100 ng of DNA. The calibration curve (*Figure 2*) is prepared with standard solutions of *lambda*-DNA (Molecular Probes, Switzerland). For each measurement 200 μL of the working solution is loaded into a microplate well and mixed with 10 μL of sample. To reduce the experimental error, all assays are performed in duplicates. The data of the DNA release is usually presented in the normalized form, $\Theta_{s/ns}$, where the DNA concentration in the supernatant of the sheared sample $c[DNA]_s$ is divided by the DNA concentration obtained from the supernatant of the not sheared sample $c[DNA]_{ns}$, at the corresponding times. This presentation of the data is chosen to emphasize the consequence of the release on any bioprocess, e.g. the DNA contamination of the culture fluid and therefore the product.

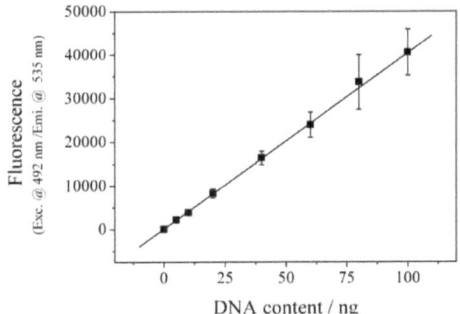

Figure 2 Calibration curve of the DNA assay. Linear correlation y = 74.01 + 403.09x.

2.2.5 Detection of Apoptotic and Necrotic Cell Death

Distinct morphological alterations, the condensation of the nuclear chromatin, the disintegration of the nuclear envelope, DNA fragmentation, and the loss in membrane asymmetry are the typical characteristics during the apoptotic process and can be used for detection (Sgonc and Gruber, 1998).

Since the particular steps of the apoptotic cascade occur in a certain sequence, the different detection techniques are applicable only at certain time points and may vary in their signal intensity. For instance, the externalization of phosphatidylserine (PS) was shown to precede nuclear changes including DNA fragmentation (Martin et al., 1995;

O'Brien et al., 1997). Based on this phenomenon, a detection method utilizing fluorescence activated cell sorting (FACS) was introduced (Andree et al., 1990; Koopman et al., 1994), capable to recognize an apoptotic subpopulation within one hour after induction (Darzynkiewicz et al., 1997). As mentioned above, due to the translocation of phosphatidylserine to the outer part of the apoptotic cells, while keeping the cell membrane integrity, apoptotic cells will selectively bind the anticoagulant annexin-V (AnnV) (Koopman et al., 1994; Rasola and Geuna, 2001). Moreover, due to the loss of intracellular water, apoptotic cells will feature a decreased forward light scattering (FSC) resulting from size reduction and an increased side scattering signal (SSC) caused by an increase of the cell surface granularity (Darzynkiewicz et al., 1992; Portier et al., 2006; Rasola and Geuna, 2001). Cell staining with fluorescein isothiocyanate (FITC)-conjugated (green fluorescence) AnnV can be performed in a Ca^{2+} enriched binding buffer and will identify apoptotic cells. Since AnnV cannot diffuse through an intact cell membrane this method allows discriminating between, apoptotic, necrotic, and viable cells, by combining the assay with a propidium iodide (PI) vital staining (red fluorescence).

For the sake of completeness other detection methods shall be briefly itemized. A well established tool for detecting an apoptotic cell population makes use of the characteristic cleavage pattern of the DNA. This phenomenon is usually analyzed by gel electrophoresis, revealing the typical DNA-ladder with fragments of the molecular weight equivalent to multiples of a nucleosome (Wyllie, 1980). Moreover, the loss of mitochondrial membrane potential (Poot and Pierce, 1999), measurement of caspase activity (Kohler et al., 2002), and microscopic inspection of the cell population can be used as powerful techniques for the detection of apoptotic cells.

The FACS analysis is performed using a Beckman Coulter Cytomics FC 500 flow cytometry system equipped with an Argon laser (wavelength 488 nm, 20 mW output) and a red Helium laser (wavelength 633 nm, 20 mW output). For determination of apoptotic, necrotic, and viable sub-populations the AnnV-FITC Apoptosis Detection Kit I (BD Bioscience Pharmingen, USA) is used. After the respective experiment, the cell solution is centrifuged, the supernatant discarded, the cell pellet washed twice in cold phosphate buffered saline (PBS), and a second centrifugation step performed. After resuspending the pellet in 1x binding buffer (provided within the kit), 100 µL of this suspension are transferred to a 15 mL falcon tube and mixed with 7 µL AnnV and 7 µL PI. To discriminate between apoptotic, necrotic, and viable cells, a parallel staining with AnnV and PI is used. The DNA-intercalating dye PI will penetrate only necrotic cells whereas AnnV will bind to both necrotic as well as apoptotic cells. Viable cells will not show any fluorescence signal. The mixture is gently vortexed and incubated for 15 minutes at room temperature in the dark. With adding another 400 μL of binding buffer and 200 μL of PBS the sample is ready for FACS analysis. During each measurement 10000 cells are counted with a throughput of approximately 50-100 cells s^{-1}.

As positive control, the culture medium is supplemented with 1 μmol Staurosporine (Deshmukh and Johnson, 2000; Zhang et al., 1998), an alkaloid isolated from *Streptomyces staurosporeus* cultures. Its chemical activity originates from the high affinity to the ATP-binding site of the kinase, preventing the phosphorylation. Similar as for sheared and unsheared cultures, samples exposed to Staurosporine where analyzed by FACS, DNA measurements, and light-microscopy. As can be seen from *Figure 3*, a significant increase of side scattering, indicating cell granularity, with a decrease of forward scattering, indicating a reduction of the cell size, was typically measured for positive controls.

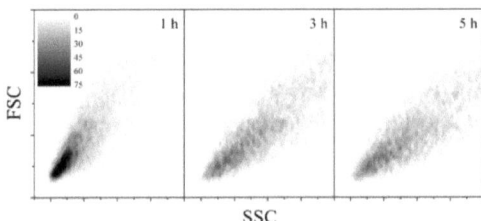

Figure 3 FACS analysis of HEK cells endued with 1 μmol of Staurosporine, x-axis: forward light scatter, y-axis: side light scatter. From left to right: after 1, 3, and 5 hours of incubation.

In comparison to the control sample without Staurosporine, indicated by the dashed line, the AnnV fluorescence for the positive control is clearly shifted to higher intensities (*Figure 4*). Moreover, it is shown that different concentrations of Staurosporine (1 or 2 μmol) do not lead to different results. In contrast to the variation in AnnV fluorescence, after treatment with Staurosporine, no PI increase could be observed as well as any release of DNA into the surrounding medium (data not shown).

Chapter 2 - Materials and Methods

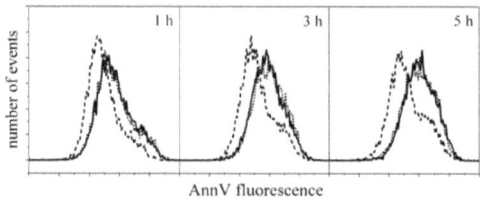

Figure 4 FACS analysis of HEK cells, AnnV fluorescence. (dashed line) negative control sample without Staurosporine; (solid line) samples endued with 1 µmol; (dotted line) samples endued with 2 µmol of Staurosporine, respectively. From left to right: after 1, 3, and 5 hours of incubation.

To visualize cells undergoing apoptosis, microscope images of the cells are taken and presented in *Figure 5*. Both cell lines show the characteristic blebbing and increased surface granularity of apoptotic cells.

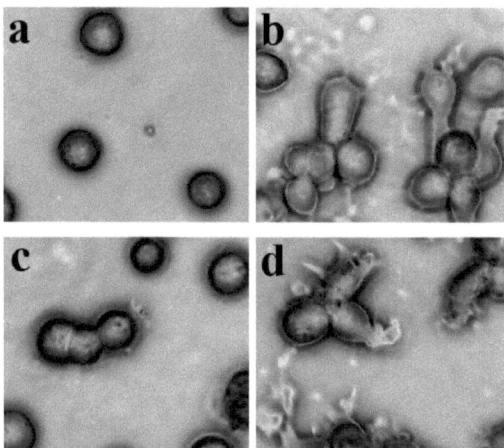

*Figure 5 Images of cells undergoing apoptosis induced by the addition of 1 µmol Staurosporine. (**a**) CHO w/o SP; (**b**) CHO w SP; (**c**) HEK w/o SP; (**d**) HEK w SP.*

The detection of necrotic cells is done by the measurement of intracellular components released into the culture medium (Croughan and Wang, 1989; Racher et al.,

Chapter 2 - Materials and Methods

1990) (see section 2.2.4), combined with microscopical inspection and FACS analysis. Necrotic cells swell and show an increased forward scatter. Moreover, PI as well as AnnV will penetrate the damaged cell membrane and both fluorescence signals amplify.

2.2.6 Bradford Protein Assay

The Bradford protein assay is based on the absorbance shift of Coomassie Brilliant Blue G-250 (Bio-Rad Laboratories, Switzerland) when bound to arginine and aromatic residues (Bradford, 1976). The anionic, bound form has an absorbance maximum at 595 nm whereas the cationic, unbound form has an absorbance maximum at 470 nm. For the total protein determination, 20 µL of sample are mixed with 1000 µL of Coomassie Brilliant Blue G-250 and incubated for 5 minutes at RT. Afterwards the absorbance at 590 nm is measured (*Figure 6*).

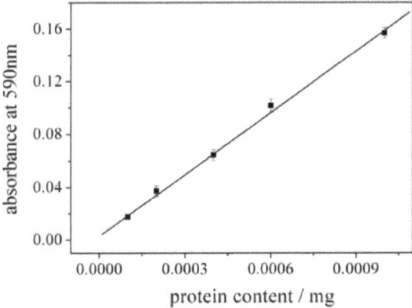

Figure 6 Calibration curve of the Bradford assay. Linear correlation y = 0.00275 + 155.91x.

The total amount of protein θ within the culture fluid serves as contamination index ϕ and can be related to the SEAP concentration at each time i:

$$\phi_i = \frac{\theta_i}{SEAP_i} \tag{5}$$

2.3 Visualization of the Flow Field

The visualization of the flow field within a reactor is performed by the use of a rheoscopic fluid, which is a suspension of microscopic crystalline platelets, able to align in dynamic currents. For this purpose, the respective flow device is installed, filled with the rheoscopic fluid AQ-RF (Kalliroscope Corp., USA) and the agitation at the respective rate started. The steady-state is usually reached within one minute and photographically recorded (Casio Computer Co., Ltd., Japan).

2.4 $k_L a$ Measurement

The oxygen transfer capacity of a bioreactor can be described and analyzed by means of the volumetric mass transfer coefficient, $k_L a$. There are several methods for the determination of the oxygen mass transfer coefficient available and a review can be found elsewhere (Van't Riet, 1979). In this work the well established dynamic gassing-out method (Bandyopadhyay et al., 1967) is used. Briefly, after adjusting the respective agitation and aeration rate, a two-point calibration (at 0 and 100% air saturation) of the dissolved oxygen sensors (Mettler-Toledo, Switzerland), is carried out. After degassing the medium by bubbling with nitrogen in order to remove all dissolved oxygen (DO), the nitrogen feed is interrupted, the air supply is opened and the evolution of the dissolved oxygen concentration recorded. The DO concentration is monitored until a new equilibrium state is reached. All experiments are performed in triplicates using water as working media, since preliminary experiments had shown a negligible difference in the mass transfer characteristics between water and the used cell culture media (data not shown).

The oxygen mass balance is given by the accumulation of oxygen per unit time equal to the rate of oxygen dissolution into water:

$$\frac{dc_{O_2}}{dt} = k_L \frac{A}{V} \left(c_{L,O_2}^{sat} - c_{L,O_2,t} \right) \tag{6}$$

where $A/V = a$ is the specific surface area, k_L is the mass transfer coefficient, and $c_{L,O_2,t}$ the dissolved oxygen concentration at time t, where c_{L,O_2}^{sat} indicates the saturation concentration. The absolute value of c_{L,O_2}^{sat} for water was taken to be 6.75 mg L^{-1} at 37° C. It is noted that this method is based on the assumptions that gas and liquid are ideally mixed, the gas composition and the pressure inside the vessel are constant, and that the liquid film resistance controls the mass transfer. By multiplying the mass transfer coefficient k_L by the specific area, a, the lumped coefficient or the process rate constant $k_L a$ is obtained. In all measurements c_{L,O_2} (at $t = 0$) is equal to 0. After integration of equation (6), one obtains the final concentration profile:

$$y(t) = \frac{c_{L,O_2,t}}{c_{L,O_2}} = 1 - \exp\{-k_L a \cdot t\}$$

(7)

$$\ln(1 - y(t)) = -k_L a \cdot t$$

By plotting $\ln(1 - y(t))$ versus time, the rate constant can be read directly from the negative slope of the graph.

2.5 Mixing Time

The determination of the mixing time is a fast and easy method to characterize the mixing efficiency of a stirring system. The mixing time is defined by the time necessary to achieve a certain, predetermined degree of homogeneity after a disturbance. Various methods for determining the mixing time, including the measurement of conductivity and pH, have been proposed (Manna, 1997). Moreover de-colorization methods (Brennan and Lehrer, 1976), and fluorescence based techniques (Einsele et al., 1978) have been developed and proven suitable.

In this work the mixing time of the reactors, t_M, is determined by a pulse technique using two identical pH-electrodes (Mettler-Toledo, Switzerland). The injection is carried out with a syringe containing 1 mL of 1.0 mol NaOH as tracer substance. In the STR the first pH probe is set 40 mm beneath the liquid surface and the second is mounted at 140 mm depth inside the reactor. The tracer is injected from the top of the reactor, 10 mm under the liquid surface. In the case of the TC and LTC units, one pH probes is mounted at either end of the reactor. The reactor is filled with water and all remaining air withdrawn. The tracer injection is performed at one end of the unit on the opposite side of the pH probe. The pH change is recorded and the time after the injection until 95 % homogeneity is defined as mixing time of the reactor. To determine the end value of homogeneity of the mixing time, the start value is set to 0 % and the final pH value is set to 100 %. All the other normalized pH values are calculated by the following equation:

$$pH_n = \frac{pH_t - pH_{t_0}}{pH_{t_e} - pH_{t_0}} \tag{8}$$

By combining the mixing time with the agitation speed, N, the dimensionless mixing time, Nt_M, representing the number of revolutions until a certain homogeneity is reached, is obtained. It approaches a constant value under turbulent conditions (Nagata, 1975).

2.6 Experimental Procedure for Hydrodynamic Stress Exposure

To investigate the effect of the hydrodynamic stress on cell death, two different devices generating either simple shear or extensional flow are used. A uniform steady simple shear flow is generated using a cup and bob rheometer (ARES rheometer, Rheometric Scientific, Germany). The inner and outer diameter of the cup and the bob are equal to 34 mm and 32 mm, respectively, leading to a gap width of 1 mm. The height of both cylinders is equal to 34 mm. For the experiments, the cell suspension (3.8 mL) is pipetted into the cup of the rheometer, the bob mounted, and the shearing with the respective value of the hydrodynamic stress started. Over the duration of the experiment several samples are withdrawn, spinned down, and the supernatant analyzed as in 2.2 described. During the experiments the temperature is maintained constant and equal to 37° C, using the jacketed cup connected to a thermostatically controlled water bath.

The second type of flow used to study cell damage is an extensional flow generated at the entrance to a sudden contraction. The set-up consisted of two syringes (Becton, Dickinson and Company, USA) connected through a capillary (Hamilton Europe, Switzerland) with an inner diameter equal to 0.20, 0.51 or 1.07 mm using a ratio of length to diameter equal to 23.5 (*Figure 7*). The flow rate is controlled using a programmable VIT-FIT syringe pump (Lambda Laboratory Instruments, Switzerland). The experiments are characterized according to the number of passes (up to 60) through the capillary and, therefore, through the region of strong elongation where the highest values of the characteristic velocity gradient (or hydrodynamic stress) are present (Blaser, 2000a; Sonntag and Russel, 1987).

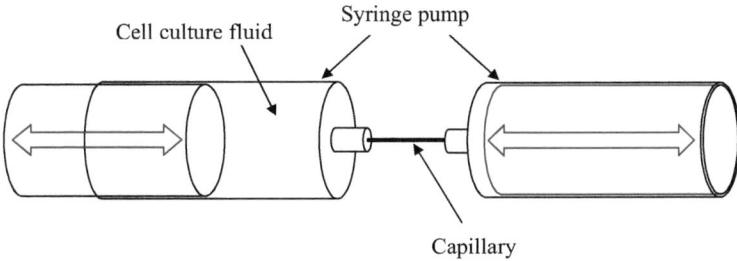

Figure 7 Set-up of the short term shearing experiment with two syringes.

For all experiments, the initial cell suspension is split into three parts. The first part is used for the shearing experiment (S), the second part as a non sheared control sample (NS) for quantifying any change of the culture over the time of the experiment, and the third part is used as positive control (PC) supplemented with 1 µmol Staurosporine (SP) (Sigma-Aldrich, Switzerland) to promote apoptosis (Deshmukh and Johnson, 2000; Zhang et al., 1998). According to observations made by McQueen et al. (1987), another parameter investigated in this study was the capillary length (i.e., the residence time). Capillaries of various relative lengths (1, 3, and 10x) but constant diameter are investigated. Care was taken to avoid the presence of air inside the syringes as well as the occurrence of cavitation. Moreover, the syringe experiments are expanded to investigate the cumulative effect of extensional flow on the cellular response. Due to possible oxygen limitations and cell settling inside the syringe, the experimental set-up described in *Figure 7*, is not suitable for sterile and long term operation, and is therefore modified as specified in *Figure 8*. The whole device is sterilized prior the experiment to avoid microbial contamination over the culture time. The reservoir is temperated to 37° C and the cell suspension slowly mixed via a magnetic stirrer. By the use of variable capillary diameter and by adjusting the flow rate, different values of the shear stress could be employed. Four different conditions are chosen, resulting in shear stresses equal to 0.03, 1.00, 10, and 170 Pa. Exposures to 0.03 Pa served as "not sheared" control. After a certain number of passes, liquid samples are withdrawn, the cell number and viability determined and the supernatant stored for further analysis at -20° C.

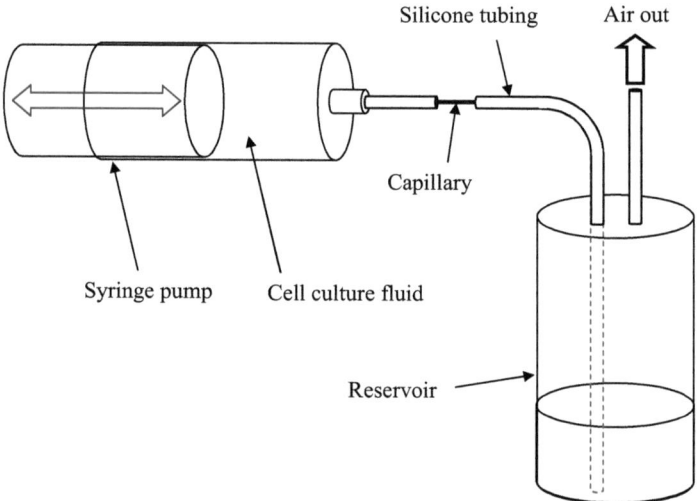

Figure 8 Set-up of the long term shearing experiment with one syringe.

Chapter 3
Response of Mammalian Cells to Environmental Stress

3.1 Introduction

3.1.1 Cell Damage due to Shear Exposure

Since mammalian cells are used as a production system for pharmaceutical proteins, researchers are studying both lethal and nonlethal effects on cells in the hydrodynamic environment of bioreactors. Concerning the response of cells exposed to shear flow, several biochemical, physiological, and culture parameters were found to be influenced, such as regrowth ability (Dunlop et al., 1994), respiration activity (Abu-Reesh and Kargi, 1989), secondary metabolite accumulation (Hooker et al., 1989), changes in cell morphology (Kretzmer and Schugerl, 1991; Ludwig et al., 1992), glucose consumption rate (Ludwig et al., 1992), release of intracellular components like DNA (Croughan and Wang, 1989; Dunlop et al., 1994; Lutkemeyer et al., 1999; Racher et al., 1990; Senger and Karim, 2003) and lactate dehydrogenase (LDH) (Abu-Reesh and Kargi, 1989; Gregoriades et al., 2000; Ludwig et al., 1992; Lutkemeyer et al., 1999; Ma et al., 2002; McQueen and Bailey, 1989; McQueen et al., 1987; Petersen et al., 1988; Racher et al., 1990; Smith et al., 1987), protein productivity (Keane et al., 2003; Motobu et al., 1998; Sun and Linden, 1999), cell viability (Dunlop et al., 1994; Hooker et al., 1989; Ludwig et al., 1992), and total cell count (Ludwig et al., 1992).

Chapter 3 - Response of Mammalian Cells to Environmental Stress

Necrosis

The most severe effect which environmental stress could cause is necrotic death, which is totally passive from the cell's point of view. An accidental traumatic event, such as hydrodynamic stress, heat, osmotic shock, or radiation (Augenstein et al., 1971; Chisti, 2001; Kretzmer, 2001) injures the cell and leads to a swelling of the cell and the organelles (Rathmell and Thompson, 1999), followed by the rupture of the plasma membrane and the disintegration of the cell. During this process the DNA is degraded into pieces of heterogeneous size. Eventually, components from the cytoplasm will be released to the extracellular environment (Abu-Reesh and Kargi, 1989; Croughan and Wang, 1989; Dunlop et al., 1994; McQueen et al., 1987) and can cause an inflammatory response within the organism.

Petersen et al. (1988) and Abu-Reesh and Kargi (1989) sheared hybridoma cells under laminar and turbulent conditions applying a Searle rheometer and found that at a critical value of 5 Pa the hydrodynamic stress leads to necrosis of the cells. Exposing a hybridoma cell line to laminar conditions in a rheometer Smith et al. (1987) observed a decrease in living cell count and an increase in LDH release, indicating necrosis, between 0.42 and 0.87 Pa. Next to mammalian cells, other systems were used to investigate cell necrosis. By measuring an increase in cell lysis and a decreasing mitochondrial activity Hooker et al. (1989) indicate a critical value of hydrodynamic stress equal to 1.3 Pa for a tobacco cell line. Spodoptera frugiperda cells, studied by Tramper et al. (1986) using a stirred tank and a rotaviscometer, started to lose viability at a hydrodynamic stress of around 1.5 Pa.

In comparison to commonly used rheometers operated under laminar conditions, Augenstein et al. (1971) introduced another type of flow device, employing capillaries of various diameters to study cell death under turbulent conditions. They observed necrotic cell death when the wall shear stress inside the capillary exceeded 200 Pa, corresponding to a Reynolds number (Re) of 4000. This experimental set-up was later adopted by McQueen et al. (1987) to investigate the effect of hydrodynamic stress on mouse myeloma cells, applying turbulent conditions ($Re < 4800$) . They reported a value of the wall shear stress equal to 180 Pa to be critical for cell lysis with its extent dependent on the capillary length. Ma et al. (2002) investigated the sensitivity of different human and insect cell lines to turbulent flow in a contractional flow device and found by means of LDH release that a hydrodynamic stress larger than 100 Pa is needed to cause cell necrosis. The shear sensitivity of myeloma and hybridoma cells was investigated by Zhang et al. (1993), employing capillaries with a sudden contraction operated under turbulent conditions with Re < 5100. Significant cell lysis was reported when the wall shear stress exceeds 240 Pa (using operating conditions as presented in the original work and corresponding to $Re \approx 3500$). Moreover, in comparison to McQueen et al. (1987), no dependency of the cell lysis rate on the capillary length (i.e. the residence time) was reported. According to their explanation the cell viscoelasticity and the short residence time in the capillary are responsible for this difference.

The significantly larger values of the critical stress observed for capillaries using turbulent conditions compared to the laminar flow in rheometers are neither in agreement with the breakage of aggregates, composed of polymer or silica particles, in extensional flow and in simple shear (Blaser, 1998; Blaser, 2000b), nor with the data of Stone et al. (1986) and Bentley et al. (1986) measuring the breakage a drops. For both systems it was shown that extensional flow is more destructive compared to simple shear flow at the same value of the characteristic velocity gradient. Moreover, by direct observation of the breakage event (Blaser, 2000a; Higashitani et al., 1991), it was shown that breakage occurs at the capillary entrance, where the rate of fluid extension is the largest, and not inside the capillary or at its exit. As all above mentioned experiments were performed at Re < 5000, combined with the fact that the flow inside the capillaries is generated by a sudden reduction of the pipe diameter, the flow at the entrance zone is most probably laminar. However, as the time required to pass through the capillary entrance is very small and inversely proportional to the velocity gradient at this point, there is a limiting time over which the cell could be deformed and eventually disrupted. Approximating the deformation of a cell to that of a drop, where a certain time is required to deform the drop before it can be broken (Bentley and Leal, 1986; Stone et al., 1986), it is reasonable to expect significantly larger values of the critical hydrodynamic stress in the extensional flow compared to simple shear flow, whereas in the first case the cells experience stress for a short time period, and in the second they are continuously deformed over considerable longer times.

Apoptosis

In addition to necrosis, cell debilitative burden, such as nutrient limitation (Mercille and Massie, 1994a; Sanfeliu and Stephanopoulos, 1999), the accumulation of toxic metabolites (Singh et al., 1994), a decreased or oscillating dissolved oxygen concentration (Mercille and Massie, 1994b), an elevated osmolality (deZengotita et al., 2002), and hydrodynamic stress (Al-Rubeai et al., 1995; Mollet et al., 2007) can lead to cell death, by triggering the apoptotic pathway.

Apoptosis, also named as programmed cell death, is a normal, ubiquitous event in the development and life of multi-cellular organisms. In contrast to necrotic death, apoptosis is a controlled and regulated process triggered by a variety of stimuli, originating either intracellular or extracellular. Intrinsic signals are produced upon the exposure to cellular stress (for instance DNA damage or oxidative stress) and initiate cell death via the mitochondrial pathway. In the other case, extrinsic stimuli, like the binding of death inducing ligands to cell surface receptors, lead to apoptosis via the death receptor pathway. A short summary of the governing pathways and the various apoptosis triggers can be found elsewhere (Laken and Leonard, 2001).

Upon receiving the apoptosis inducing signal, a series of biochemical events are initiated, causing a number of characteristic and profound changes within the cell. Morphological, cell shrinkage due to dehydration, an increased cell granularity, reflecting the different cytoplasm and nucleus consistency, and intensive cell ruffling and blebbing, can be observed (Dive et al., 1992; Franek and Fussenegger, 2005; Petit et al., 1995; Portier et al., 2006; Schmid et al., 1994). Moreover, during the caspase cascade nuclear proteins such as DNA repair enzymes are deactivated and degradative enzymes such as endonucleases are activated. These events lead to the cleavage of the DNA at the linker section into fragments of the molecular weight equivalent to multiples of a nucleosome (180 base pairs) (Gorczyca et al., 1993; Wyllie, 1980). As opposed to the internal collapse of the apoptotic cell, the outer plasma membrane will remain unchallenged and maintain its structural integrity. However, the phospholipid exposure asymmetry will be lost and phosphatidylserine (PS) originally facing the inner part of the cell will be externalized and therefore exposed on the outer side of the cellular membrane. This process can be detected as a very early event during apoptosis (Fadok et al., 1992; Koopman et al., 1994; van Engeland et al., 1998; Vermes et al., 1995). The very last event of the apoptotic pathway is the fragmentation of the cell into apoptotic bodies, which become later phagocytized (Bennett et al., 1995; Saraste and Pulkki, 2000).

As it was shown by Aloi and Cherry (1994), Sf-9 insect cells exposed to hydrodynamic stress exhibit an elevated intracellular calcium concentration, known as an early event in apoptosis preceding cell death (Trump and Berezesky, 1995). The increase was found to be almost linearly dependent on the rate of energy dissipation in the capillary, independently of the flow regime (laminar or turbulent with $500 < Re < 6000$) and residence time (capillary length). In addition to this study, various authors (Al-Rubeai et al., 1995; Aloi and Cherry, 1994; Apenberg et al., 2003; Shive et al., 2000) report inductive as well as the inhibitory effects of flow on cell apoptosis.

Use of Cell Protective Agents to Reduce Cell Damage

Surfactants such as polyethylene glycol, polyvinyl alcohol, and especially Pluronic F-68 (F-68) are routinely used in mammalian cell culture processes to protect the suspended cells from detrimental effects of shear stress originating from bubble aeration (Chattopadhyay et al., 1995; Kunas and Papoutsakis, 1989; Ma et al., 2004; Michaels and Papoutsakis, 1991; Murhammer and Goochee, 1990). Here, it shall be noted that the obtained values of the critical hydrodynamic stress mentioned in section 3.1.1 have to be handled carefully, since most of the experiments were performed with culture medium supplemented with fetal calf serum or other cell protective agents.

F-68 is a water-soluble, synthetic, nonionic block co-polymer of hydrophilic poly(ethylene oxide) and hydrophobic poly(propylene oxide) terminating in primary hydroxyl groups. The monomers are linked together in the sequence PEO_a-PPO_b-PEO_c and its average molecular weight is 8400 g mol^{-1}. For Pluronic F-68 the indices are equal to a = 75, b = 30 and c = 75 (Ahmed et al., 2001).

In literature, three protection mechanisms of F-68 are suggested: (1) the surface active substances reduce bubble bursting, film drainage and stabilize the foam (Bavarian et al., 1991; Handacorrigan et al., 1989; Ma et al., 2004), (2) cell-bubble interactions are prevented by making the rising bubbles slippery and avoiding cells rising to the surface (Garciabriones and Chalmers, 1994; Jordan et al., 1994; Michaels et al., 1995), and (3) the surfactant is anchoring at the lipid-bilayer cell membrane creating a "secondary cell wall" and increasing the flexibility of the cell membrane (Murhammer and Goochee, 1988; Murhammer and Goochee, 1990; Ramirez and Mutharasan, 1990).

Next to the cell protective effect, F-68 was shown to alter the cell's capacity for recombinant protein production (Palomares et al., 2000) and cell growth, independently of the shear environment (Alrubeai et al., 1992; Bentley et al., 1989; Mizrahi, 1975). Moreover, cell protective agents exhibit also major drawbacks connected to a decreased mass transfer (Moreira et al., 1995; Murhammer and Pfalzgraf, 1992) and additional problems in the downstream purification process (Heath and Kiss, 2007), like a decreased protein binding capacity of chromatographic resins (Low et al., 2007).

3.1.2 Flow Devices and Fluid Flow Characterization

Laminar Conditions

To investigate the effect of the flow type on cell death, two different devices generating either simple shear or extensional flow are used. Due to the rotation of the outer cylinder and the narrow gap size, the flow within the gap is always laminar with a linear velocity profile in the tangential direction (Bird et al., 2002; Rheometric Scientific Inc, 1998). Assuming the cell to be spherical, the maximum hydrodynamic stress, τ, to which it will be exposed in simple shear flow is calculated as following (Blaser, 1998):

$$\tau = \frac{5}{4}\eta\gamma \qquad (9)$$

where η is the dynamic viscosity of the culture medium (approximated to be equal to that of water) and γ the shear rate.

The second type of flow used to study cell damage under laminar conditions is an extensional flow generated at the entrance of a sudden contraction. To obtain the characteristic velocity gradient and consequently the hydrodynamic stress in extensional flow it is necessary to solve the mass and momentum balance equations for the particular operating conditions. The general form of the momentum conservation equation is taking into account the fluctuation of the local velocity due to turbulence and can be written as (FLUENT 6.2, 2005; Pope, 2000):

$$\rho \frac{D\langle U_i \rangle}{Dt} = \frac{\partial}{\partial x_i}\left[\eta\left(\frac{\partial \langle U_i \rangle}{\partial x_j} + \frac{\partial \langle U_j \rangle}{\partial x_i}\right) - \langle p \rangle \delta_{ij} - \rho\langle u_i u_j \rangle\right] \qquad (10)$$

where the first component in square brackets represents the viscose stress due to the mean velocity gradient, the second term represents the isotropic stress from the mean pressure, and the third one corresponds to the stress arising from fluctuations of the velocity field called Reynolds stresses.

When the flow in the whole domain is laminar, which is assumed to be the case when $Re_{\text{capillary}} \leq 700$, the momentum conservation equation, Eq. (10), can be solved directly since its last term can be neglected. Under these conditions, the velocity gradient tensor can be decomposed into symmetric and anti-symmetric parts. Here, the anti-symmetric part represents the rate-of-rotation tensor, Ω, which contributes to the solid body rotation. The symmetric part corresponds to the rate-of-strain tensor, E (Pope, 2000), causing cell deformation. Due to the axial symmetry of the used geometry, the corresponding components of the rate-of-strain tensor, e_{ij}, in 2-D cylindrical coordinates (r,θ,z) are as follows:

$$\begin{aligned}
e_{11} &= \partial U_r/\partial r \\
e_{22} &= U_r/r \\
e_{33} &= \partial U_z/\partial z \\
e_{13} &= e_{31} = (1/2)(\partial U_r/\partial z + \partial U_z/\partial r) \\
e_{12} &= e_{21} = e_{23} = e_{32} = 0
\end{aligned} \quad (11)$$

where e.g. U_r represents the radial velocity component. As can be found elsewhere (Bird et al., 2002; Pope, 2000), applying a rotation of the coordination system, reduces E to a diagonal form with diagonal components equal to the eigenvalues of E. For this particular case the corresponding eigenvalues of the rate-of-strain tensor can be obtained analytically and are equal to:

$$\alpha_L > \beta_L > \gamma_L \begin{cases} = \frac{1}{2}e_{11} + \frac{1}{2}e_{33} + \frac{1}{2}\left(e_{11}^2 + 4e_{13}^2 + e_{33}^2 - 2e_{11}e_{33}\right)^{0.5} \\ = e_{22} \\ = \frac{1}{2}e_{11} + \frac{1}{2}e_{33} - \frac{1}{2}\left(e_{11}^2 + 4e_{13}^2 + e_{33}^2 - 2e_{11}e_{33}\right)^{0.5} \end{cases} \quad (12)$$

Consequently, the maximum positive eigenvalue, α_L, calculated from the components of the rate-of-strain tensor, E, is used as the characteristic velocity gradient (Blaser, 1998). Therefore, the maximum hydrodynamic stress to which the cells (approximated again as spheres) are exposed in the extensional flow is evaluated as following:

Chapter 3 - Response of Mammalian Cells to Environmental Stress

$$\tau_L = \frac{5}{2}\eta\alpha_L \qquad (13)$$

where α is the characteristic velocity gradient equal to the maximum positive eigenvalue of the mean velocity rate of strain, α_L.

Turbulent Conditions

With increasing the flow rate, the flow in the capillary as well as near its exit becomes transitional or even turbulent. In comparison to laminar conditions, where the kinetic energy is dissipated due to the mean flow, in the case of turbulent conditions, there is an additional component which contributes to the dissipation of the kinetic energy, coming from the velocity fluctuation, the so-called Reynolds stresses.

These gradients are related to the local turbulent energy dissipation rate, ε, which for a Newtonian fluid is defined as (Pope, 2000):

$$\varepsilon = 2\nu \langle s_{ij} s_{ij} \rangle \qquad (14)$$

where ν represents the kinematic viscosity and s_{ij} the fluctuating rate-of-strain tensor which components are defined according to:

$$s_{ij} = \frac{1}{2}\left(\frac{\partial u_i}{\partial x_j} + \frac{\partial u_j}{\partial x_i}\right) \qquad (15)$$

As for the laminar conditions, the maximum positive eigenvalue is used to characterize the flow field, which is by definition independent on the direction of flow, also in the case of turbulence a similar approach is used. Accordingly, when the fluctuating rate-of-strain tensor contains only diagonal components (eigenvalues), the turbulent energy dissipation rate ε from Eq. (14) can be written as following:

$$\varepsilon = 2\nu\left(\alpha_T^2 + \beta_T^2 + \gamma_T^2\right) \qquad (16)$$

Using the results of the direct numerical simulation of the isotropic turbulence (Vedula et al., 2001), where it was found that the ratio α_T / β_T is equal to 2.7, and assuming the liquid to be incompressible for which $(\alpha_T + \beta_T + \gamma_T = 0)$, ε in Eq. (16) can be written as a function of the maximum positive eigenvalue of the fluctuating rate-of-strain tensor as following:

$$\varepsilon = 6\nu\alpha_T^2 \qquad (17)$$

Chapter 3 - Response of Mammalian Cells to Environmental Stress

Consequently Eq. (17) is used to evaluate the maximum positive eigenvalue of the fluctuating rate-of-strain tensor, α_T, using ε, calculated from Computational Fluid Dynamics (CFD). To relate the response of the cells to the hydrodynamic stress, it is essential to compare their size to the characteristic length scale of the flow. When the cell size is significantly smaller than the size of the smallest turbulent eddies, characterized by the Kolmogorov length scale (Pope, 2000):

$$\kappa = \left(\frac{v^3}{\varepsilon}\right)^{1/4} \qquad (18)$$

any cell damage is solely controlled by the local hydrodynamics within an eddy. Due to the nature of the turbulence, the eddy size as evaluated through Eq. (18) should be understood as an approximation rather than an absolute value. The corresponding hydrodynamic stress, τ_T, to which the cell is exposed under turbulent conditions can be estimated using Eq. (13), applying the maximum positive eigenvalue of the fluctuating rate of strain, α_T, resulting in:

$$\tau_T = \frac{5}{2}\eta \alpha_T \qquad (19)$$

Another situation arises when κ is significantly smaller than the cell size, and therefore, the cell resides in the inertial subrange. Now, the hydrodynamic stress to which the cell is exposed, also called dynamic pressure (Hinze, 1955), results from the difference in the pressure between two points separated by the distance D and is equal to:

$$\tau = \frac{1}{2}\rho u'^2(D) \qquad (20)$$

where ρ is the fluid density and u'^2 the root-mean-square velocity fluctuation which is equal to (Pope, 2000):

$$u'^2(D) = C_2(\varepsilon D)^{2/3} \qquad (21)$$

with the constant C_2 approximately equal to 2 (Pope, 2000). Substituting Eq. (21) into Eq. (20) and assuming that the separation distance, D, is comparable to the cell size, d_{cell}, it is possible to approximate the hydrodynamic stress to which cells are exposed when their size is in the inertial subrange as follows:

$$\tau_{is} = \rho(\varepsilon d_{cell})^{2/3} \qquad (22)$$

τ_{is} is calculated by assuming an undeformed cell shape with a diameter equal to the mean value measured by the CASY counter, and therefore, the obtained values serve as an approximation of the real values.

Numerical Details

To obtain all above mentioned quantities the flow through the capillaries is characterized by Computational Fluid Dynamics using the commercial software Fluent v6.2 and applying the density and viscosity of water. In dependency of the experimental set-up values for 37° C (ρ = 993 kg m^{-3}, μ = 0.697 mPa s) or 22° C (ρ = 998 kg m^{-3}, μ = 0.955 mPa s) were used. Due to the flow symmetry along the *x*-axis, a 2-D axisymmetric mesh is used to calculate the flow field within the used geometry. For $Re \leq 700$, the solution is obtained using a laminar solver, while for $Re > 700$ the contribution of the turbulence (last term on right-hand-side of Eq. (10)) is included using a Reynolds-Stress-Model (RSM) (FLUENT 6.2, 2005), where the transport equations for individual components of the Reynolds stress tensor, $\langle u_i u_j \rangle$, are directly solved. Since the turbulence in the capillary entrance does not adjust fast enough to the mean velocity gradient, as confirmed by the large ratio between the strain rate and the turbulence decay rate $\alpha_L k/\varepsilon \approx 100$, it is appropriate to use the RSM to describe the flow (Pope, 2000). To properly resolve the region near the walls a two-layer model (FLUENT 6.2, 2005) was adopted. For both conditions, laminar as well as turbulent, second order spatial discretization of all variables and pressure-velocity coupling using a SIMPLE scheme (FLUENT 6.2, 2005), is used.

3.2 Results and Discussion

3.2.1 Cell Response to Laminar Simple Shear Flow

According to the method described (2.6), both cell lines are exposed to various magnitudes of hydrodynamic stress using the rheometer. An example of the AnnV and PI fluorescence, measured for unsheared and sheared samples of HEK cells, after 4 hours applying a hydrodynamic stress equal to 0.59 and 2.09 Pa is presented in *Figure 9*. The corresponding analysis for CHO cells is given in *Figure 10*.

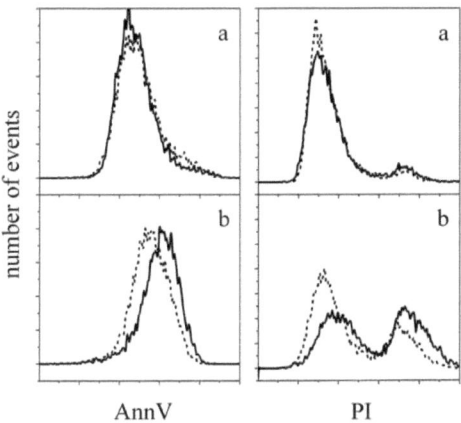

Figure 9 Example of FACS analysis of HEK cells, PI and AnnV fluorescence. (---) unsheared control sample; (—) sheared sample after 4 hours applying a hydrodynamic stress equal to (a) 0.59 Pa and (b) 2.09 Pa.

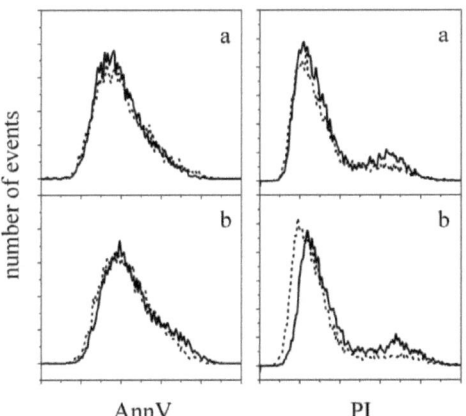

Figure 10 Example of FACS analysis of CHO cells, PI and AnnV fluorescence. (---) unsheared control sample; (—) sheared sample after 4 hours applying a hydrodynamic stress equal to (a) 0.59 Pa and (b) 2.09 Pa.

It can be seen, that exposure to a hydrodynamic stress of 0.59 Pa will not lead to an increase in AnnV and PI fluorescence. On the other hand, when the hydrodynamic stress is increased to 2.09 Pa, the sheared culture shows a significant increase in both fluorescence signals, when compared to the unsheared control culture. According to the description of the FACS analysis, this indicates that for HEK cells a hydrodynamic stress equal to 0.59 Pa is below the threshold while a hydrodynamic stress of 2.09 Pa leads to necrosis. For the CHO cell line the FACS analysis revealed an attenuated, but similar trend (*Figure 10*).

To clarify this observation and verify necrotic cell death, microscope images of unsheared and sheared cells applying a hydrodynamic stress of 2.09 Pa are taken and shown in *Figure 11* and *Figure 12*. Viable cells show a uniform circular shape, with a bright halo around the cell membrane, whereas the population of sheared cells contains mostly cell debris with only very few intact cells remaining, indicating massive cell damage as a result of the applied flow.

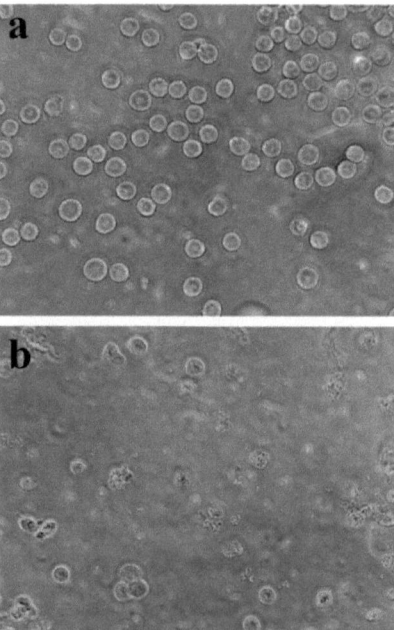

Figure 11 Images of CHO cells. (a) unsheared control sample; (b) sheared sample after applying a hydrodynamic stress equal to 2.09 Pa over 4 h using the cup and bob rheometer.

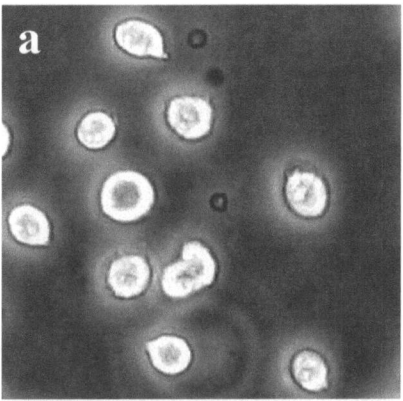
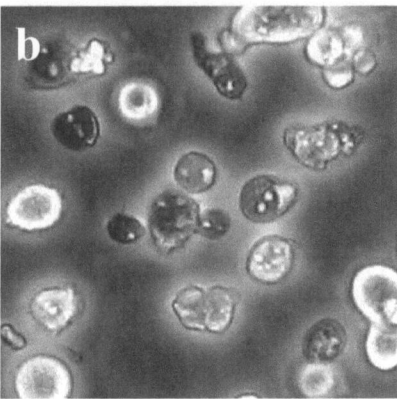

Figure 12 Images of HEK cells. (a) unsheared control sample; (b) sheared sample after applying a hydrodynamic stress equal to 2.09 Pa over 4 h using the cup and bob rheometer.

Combining these observations together with those shown in section 2.2.5 it is clear that FACS analysis serves as a good qualitative tool to distinguish between apoptosis or necrosis but can hardly quantify the amount of necrotic cells, in particular when the kinetics of cell death is of main interest. Therefore, the direct measurement of the DNA content in the supernatant is used in the following as additional detection method for necrotic cell death. Unless the cell is viable and its membrane intact, intracellular molecules like DNA will not pass into the culture media and therefore the respective concentration in the supernatant and consequently also the ratio $\Theta_{s/ns}$ will be constant. An example of the time evolution of $\Theta_{s/ns}$ for HEK and CHO cells applying various values of hydrodynamic stress is shown in *Figure 13*. The obtained data can be split into two groups. The first group contains data obtained for HEK cells using τ below 1.19 Pa and for CHO cells using τ below 0.59 Pa. As it can be seen, $\Theta_{s/ns}$ over time follows the same kinetic with no dependency on the applied hydrodynamic stress and the final DNA amount in the supernatant for the sheared sample is only slightly higher than in the unsheared control sample. This can be explained by the fact that under these conditions the viable cells are not affected by the flow and there is no DNA release from this population. However, there are some dead cells within the population (starting viability was around 90 % ± 5 %) at the beginning of the experiment and the measured DNA in the supernatant is most probably coming from these already dead cells. In the control sample this release process is mostly diffusion limited while in the sheared sample the mixing accelerates this transport. This leads to a faster permeation of the DNA from the dead cells to the supernatant and, therefore, to slightly higher transport rates for the sheared compared to the unsheared samples (*Figure 13*).

The second group contains data where the hydrodynamic stress is above the threshold value and therefore also DNA release from viable cells will contribute to the increase of $\Theta_{s/ns}$, leading to faster release kinetics and higher plateau values of $\Theta_{s/ns}$. After the fast initial disruption of the weakest cells within the population, a slower, diffusion controlled process, leads to the slight increase of $\Theta_{s/ns}$ as observed in *Figure 13*. Considering that the bursting or membrane tension is independent of the cell size (Born et al., 1992) resulting in a counteracting pressure being inversely proportional to the cell size (Hinze, 1955), the different plateau values are also related to the cell size distribution of the sample. In other words, the smaller the cell the greater the surface force counteracting the deformation. Therefore, by applying a certain hydrodynamic stress, only a fraction of cells withstand the pressure and survive. Similar observations, although at higher threshold values, were made for plant (Dunlop et al., 1994) and hybridoma cells (Born et al., 1992). In both cases, a fast initial cell death precedes a secondary damage at a lower rate leveling in different steady state values. In the case of plant cells this effect was observed for various biological parameters like the cells membrane integrity, the mitochondrial activity, and the regrowth ability, whereas for hybridoma cells only the total and viable cell concentration was investigated.

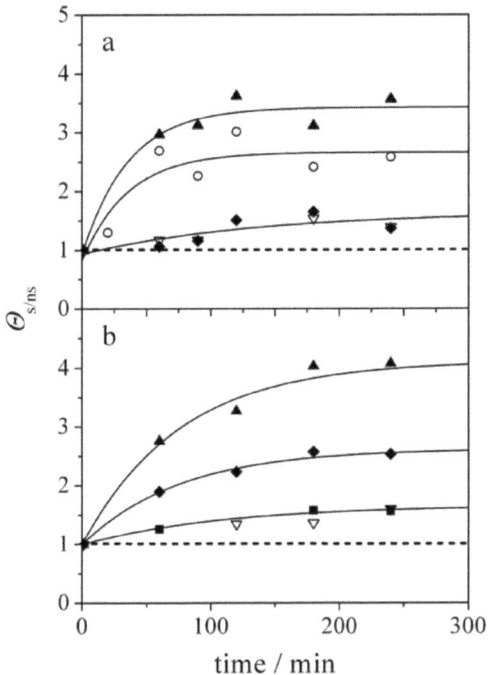

Figure 13 Example of DNA release upon simple shear exposure over the period of 4 hours. (a) HEK cells (b) CHO cells. (—) exponential fit (▲) 2.09 Pa (○) 1.67 Pa (♦) 1.19 Pa (■) 0.59 Pa (▽) 0.12 Pa.

In summary the obtained results of the DNA release as a function of the applied hydrodynamic stress are presented in *Figure 14*. It can be seen that the threshold value of the hydrodynamic stress τ_{crit} for CHO cells is between 0.59 and 1.19 Pa and between 1.19 and 1.67 Pa for HEK cells, as also measured by the FACS analysis. This is in good agreement with values reported in the literature ranging from 0.8 to 5 Pa (Abu-Reesh and Kargi, 1989; Hooker et al., 1989; Petersen et al., 1988; Smith et al., 1987; Tramper et al., 1986). Since it is known that Pluronic F-68 is used as a protective agent, reducing the hydrodynamic effect of flow on cells, one would expect a reduction in $\Theta_{s/ns}$ and

Chapter 3 - Response of Mammalian Cells to Environmental Stress

consequently an increase of the threshold value of the hydrodynamic stress, when supplementing the culture media with F-68.

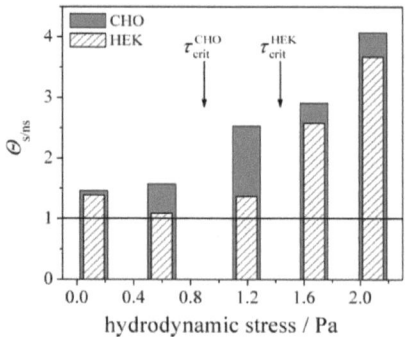

Figure 14 Average $\Theta_{s/ns}$ from CHO and HEK cells upon shear exposure for 4 hours, measured for various values of the hydrodynamic stress (exp. error within 18 %).

Therefore the shear experiments are repeated with culture medium supplemented with 1 % w/v F-68. After 4 hours of shearing applying a hydrodynamic stress equal to 2.09 Pa, which is above the threshold value for both cell lines, $\Theta_{s/ns}$ reduces by a factor of 1.5 for HEK and 1.9 for CHO cells, respectively (data not shown). Assuming a similar increase of $\Theta_{s/ns}$ with the hydrodynamic stress as observed in *Figure 14*, the addition of F-68 would lead to an increase of the critical value of hydrodynamic stress causing necrosis by 30 to 70 % for HEK and CHO cells respectively.

Cell death caused by apoptosis is not observed under the simple shear conditions investigated in this work. Possible reasons could be that under the applied conditions this mechanism is not present at all or because of the experimental set-up, which does not allow to reach either a time scale short enough (without unpredictable side effects like acceleration of the cup) or large enough (due to microbial contamination) to trigger apoptosis.

3.2.2 Extensional Flow - Flow Field Characterization

The extensional flow field is characterized using the CFD software Fluent v 6.2. The CFD simulations were validated by comparing the predicted values of the pressure drop with values obtained analytically (Perry, 1997), see *Table 2*. For all conditions good agreement within 12 % error, is found.

As described in section 3.1.2, for laminar conditions the hydrodynamic stress, τ, to which the cells are exposed, is calculated using the maximum positive eigenvalue of the mean velocity rate of strain tensor, α_L, in Eq. (13).

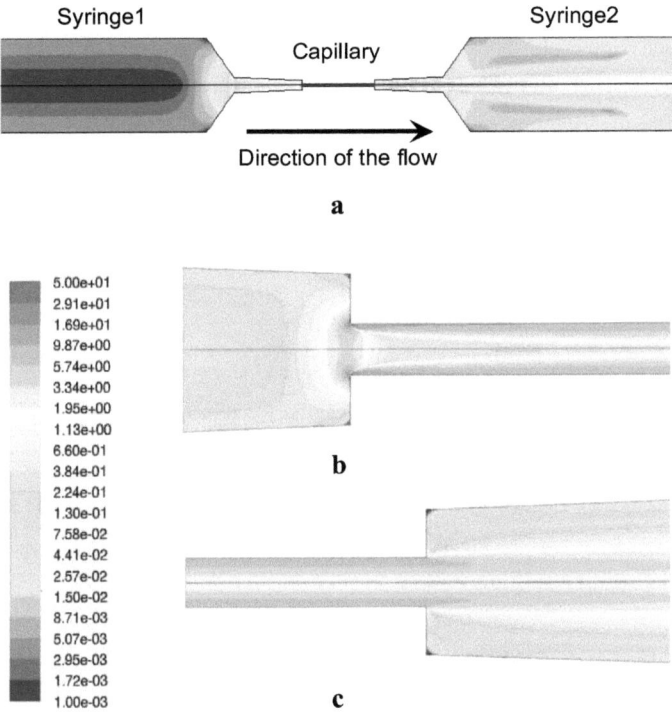

Figure 15 Contour plot of the hydrodynamic stress as evaluated from Eq. (13) in the (a) complete geometry used for CFD simulation together with details of the capillary (b) entrance and (c) exit. Diameter of the capillary $D_{capillary} = 0.51$ mm, flow rate $Q = 5.94$ mL min^{-1}.

An example of the contour plot of τ in the whole domain used for the simulation, together with details of the capillary entrance and exit, calculated for laminar conditions ($Re_{capillary}$ = 258) using an inner capillary diameter equal to 0.51 mm with a length of 11.9 mm, is presented in *Figure 15a-c*. Due to the nature of the axi-symmetric extensional flow, the values of τ_L at the capillary entrance are approximately two to three orders of magnitude larger than those in the bulk. After a certain capillary length the velocity profile approaches that in a pipe (Bird et al., 2002; Perry, 1997), with the largest values of the velocity gradient near to the capillary wall (*Figure 15b*). As the liquid leaves the capillary, it forms a jet stream where the values of the velocity gradient slowly decay to negligible values (*Figure 15c*). For comparison, values of τ_L along several streamlines at various radial positions are presented in *Figure 16*. The stream function ψ is defined as (Bird et al., 2002):

$$U_z = -\frac{1}{r}\frac{\partial \psi}{\partial r}, U_r = \frac{1}{r}\frac{\partial \psi}{\partial z} \qquad (23)$$

where ψ is made dimensionless by dividing it by $U_{capillary} r^2_{capillary}$. Hence, $\Psi = 0$ and 0.5 corresponds to the streamline along the centerline and along the wall, respectively. As can be seen, the largest values of τ_L occur at the capillary entrance and increase with increasing radial distance from the axis.

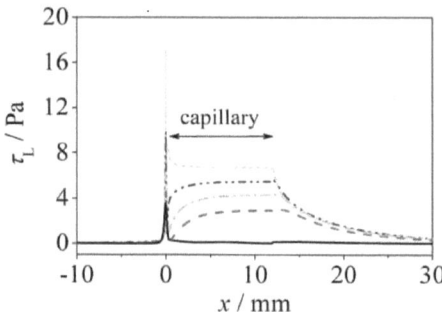

Figure 16 Hydrodynamic stress as evaluated from Eq. (13) along various streamlines (Ψ equal to 0 (—), 0.1 (– –), 0.2 (– · –), 0.3 (– · · –) and 0.4 (- -) ($\Psi = 0$ represent the capillary axis and $\Psi = 0.5$ the capillary wall; x = 0 mm represents the capillary entrance and x = 11.9 mm the capillary exit). Diameter of the capillary $D_{capillary}$ = 0.51 mm, flow rate Q = 5.94 mL min^{-1}.

Chapter 3 - Response of Mammalian Cells to Environmental Stress

To emphasize the effect of the capillary entrance with respect to the region near the wall, it shall be noted that the flow near the wall is similar to simple shear flow with hydrodynamic stress values being ½ of those for extensional flow (see Eq. (9) and (13)). This would mean that even at the same velocity gradient at the capillary entrance and near the wall, the hydrodynamic stress to which the cells are exposed is larger at the capillary entrance. Therefore, based on the simulations, whenever the flow is laminar the hydrodynamic stress is evaluated through Eq. (13).

Applying higher flow rates, the flow in the capillary becomes transitional or weakly turbulent ($Re_{capillary}$ > 700). If the cell death will be affected by turbulence, one would expect that either the hydrodynamic stress originating from the local gradients of the fluctuating velocity, τ_T, evaluated from Eq. (13) using α_T from Eq. (17) (i.e., the cell size is below the Kolmogorov length scale, Eq. (18)), or the hydrodynamic stress coming from differences in the velocity fluctuations over the cell diameter, τ_{is}, evaluated from Eq. (22) (i.e., the cell size is above the Kolmogorov length scale, Eq. (18)) will be significantly larger compared to the hydrodynamic stress due to the mean velocity gradient (Eq. (13) using α_L). An example of the above mentioned hydrodynamic stress values along various streamlines for $Re_{capillary}$ ≈ 2400 is presented in *Figure 17a-c*. It can be seen that for both laminar as well as weakly turbulent conditions the effect of the mean velocity gradient is largest at the capillary entrance, see *Figure 17a*. Since turbulence does not adjust as fast as the mean velocity, values of τ_T at the capillary entrance at corresponding radial positions are smaller than those of τ_L (*Figure 17b*). Significantly larger values of τ_T near the wall ($\Psi = 0.4$) with respect to the capillary axis ($\Psi = 0$) are related to the development of the boundary layer with largest values of the energy dissipation rate at the pipe wall.

The capillary exit is characterized by the formation of a turbulent jet, where a significant portion of energy is dissipated due to turbulence, leading to large values of α_T (and consequently τ_T). Small variation of the α_T at different radial positions leads to a relatively narrow distribution of τ_T across this jet with values varying by a factor of approximately 2, which is significantly smaller than at the capillary entrance. Finally, in *Figure 17c* values of τ_{is} as calculated through Eq. 22 using a cell diameter equal to the average value of the undeformed cell (11 µm) are presented. Considering the maximum and

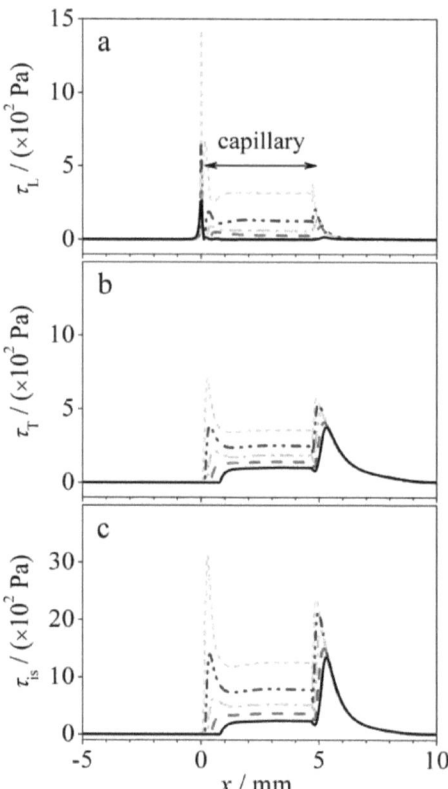

Figure 17 Comparison of the hydrodynamic stresses originating from the mean velocity gradient (a), from turbulent velocity fluctuation (b), and from inertial subrange (c) along various streamlines (Ψ equal to 0 (—), 0.1 (– –), 0.2 (– · –), 0.3 (– · · –) and 0.4 (- -) where $\Psi = 0$ represent the capillary axis and $\Psi = 0.5$ the capillary wall). Diameter of the capillary $D_{capillary} = 0.2$ mm, flow rate $Q = 21.6$ mL min^{-1}. Note the different scale used for the stresses.

the minimum cell size, a variation in τ_{is} by ± 30 % around the averaged value is observed. As it can be seen values of τ_{is} are, for this particular flow conditions, the largest and therefore one would expect that cell damage will be due to the dynamic pressure.

Due to the radial variation of the hydrodynamic stress for all cases and different locations of the largest values of the individual hydrodynamic stresses (either at the capillary entrance or its exit), their mass-weighted average values τ_i^{ave} with $i = L, T,$ is calculated over the appropriate surfaces (FLUENT 6.2, 2005), are used to characterize each flow condition. These surfaces are constructed by connecting the maxima of the individual hydrodynamic stresses at different radial positions. In *Figure 18* an example of τ_{is} together with the constructed surface (black line) at the capillary exit is illustrated. The comparison of all mentioned hydrodynamic stresses which could cause damage to the cells is presented in *Table 2*.

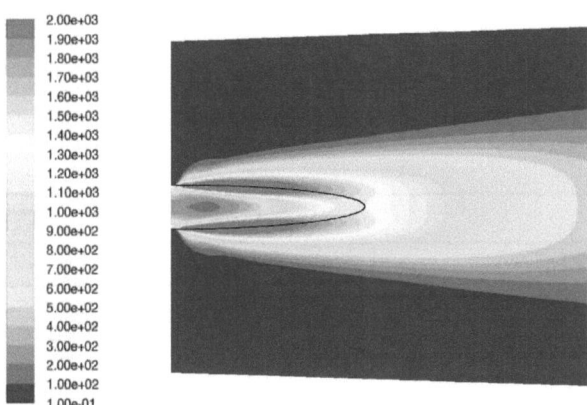

Figure 18 Example of the surface (black line) used to calculate the mass-weighted average values of the hydrodynamic stresses τ_T and τ_{is}. Background: contour plot of τ_{is} calculated for the flow rate $Q = 21.6$ mL min^{-1} and a capillary diameter equal to 0.2 mm using $d_{cell} = 11$ μm.

As it can be seen, when the flow is laminar the hydrodynamic stress arising from the mean velocity gradient located at the capillary entrance dominates and any damage to the cells is related to τ_L. On the other hand, when the flow becomes weakly turbulent, $Re_{\text{capillary}}$ around 1300, and due to the large cell size with respect to the Kolmogorov length scale, κ, the dynamic pressure over the cell size in the turbulent jet at the capillary exit becomes the largest from all investigated stresses. In the following analysis depending on the operating conditions the largest value of the hydrodynamic stress from *Table 2* will be used.

Q	$U_{capillary}$	Re	τ_{wall}	τ_L^{ave}	τ_T^{ave}	τ_{is}^{ave}	Δp_{calc}	Δp_{CFD}
			$D_{capillary} = 0.20$ mm					
mL min^{-1}	m s^{-1}		Pa	Pa	Pa	Pa	Pa	Pa
3.56	1.888	395	72	**105**	–	–	10637	10952
7.14	3.788	792	145	**237**	85	185	29808	29669
12.17	6.454	1349	247	**444**	203	591	72391	71866
16.00	8.488	1774	324	611	303	**1007**	115603	116679
21.60	11.459	2395	524	836	453	**1719**	209271	206482
26.97	14.310	2991	869	1029	616	**2588**	339115	316161
			$D_{capillary} = 0.51$ mm					
Q	$U_{capillary}$	Re	τ_{wall}	τ_L^{ave}	τ_T^{ave}	τ_{is}^{ave}	Δp_{calc}	Δp_{CFD}
mL min^{-1}	m s^{-1}		Pa	Pa	Pa	Pa	Pa	Pa
1.79	0.146	78	2.18	**2.00**	–	–	228	231
2.37	0.194	103	2.90	**3.05**	–	–	311	318
5.94	0.484	258	7.26	**9.22**	–	–	917	958
			$D_{capillary} = 1.07$ mm					
Q	$U_{capillary}$	Re	τ_{wall}	τ_L^{ave}	τ_T^{ave}	τ_{is}^{ave}	Δp_{calc}	Δp_{CFD}
mL min^{-1}	m s^{-1}		Pa	Pa	Pa	Pa	Pa	Pa
5.94	0.110	123	0.79	**0.49**	–	–	84	90
10.71	0.199	222	1.42	**1.09**	–	–	162	182

Table 2 Fluid flow characteristics of the sudden contraction as calculated from CFD. Characteristic values of the hydrodynamic stress are presented in bold.

3.2.3 Cell Response to Extensional Flow (Short Term)

To investigate the influence of extensional flow on cell death, the cultures are exposed to various magnitudes of hydrodynamic stress and number of passes through a capillary. In addition, based on the observation that near the wall the velocity gradient could reach comparable values to those at the capillary entrance, combined with the experimental data of McQueen et al. (1987), where under turbulent conditions an effect of the capillary length (e.g. the residence time) on the cell viability was observed, it was decided to use also capillaries with various lengths.

Figure 19 shows a plot of the measured forward (FSC) and side (SSC) scattering (plotted versus each other) of sheared and unsheared control samples of HEK cells applying average hydrodynamic stresses, τ_L^{ave}, in the range from 0.49 to 9.22 Pa. Under these conditions the flow is laminar through the whole domain and according to the section above, the capillary entrance is the area of highest hydrodynamic stress. As can be seen from *Figure 19* by comparing the unsheared samples (left column) and the sheared one (right column), no significant change is observed when the average hydrodynamic stress does not exceed a value of 1.09 Pa. Applying larger hydrodynamic stress leads to an increase of cell granularity, indicated by the increase of side-scattering, and a decrease in cell size as illustrated by the decrease of forward scattering. By microscopic inspection of the cells and DNA measurements (data not shown) in the supernatant, it is found that even the highest value of the average hydrodynamic stress presented in *Figure 19* after applying 60 passes through the capillary does not lead to any cell membrane damage.

This observation differs significantly to the conclusions made under simple shear flow (*Figure 13*), where under such hydrodynamic stresses severe necrosis is observed. On the other hand, as can be seen in *Figure 20*, the cell shape exhibits typical features of a cell undergoing apoptosis (see section 2.2.5).

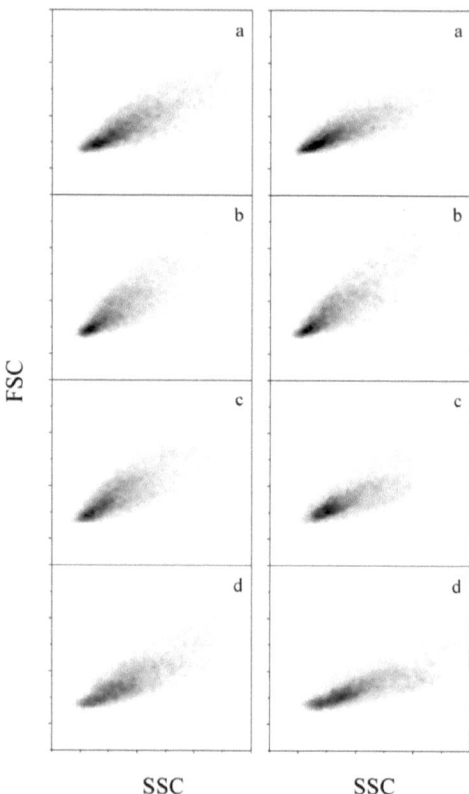

*Figure 19 FACS analysis of sheared HEK cells. x-axis: forward scatter, y-axis: side scatter. Left column: unsheared control, right column: sheared samples with an applied shear stress equal to (**a**) 0.49 Pa (**b**) 1.09 Pa (**c**) 3.05 Pa (**d**) 9.22 Pa.*

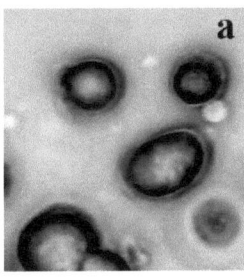 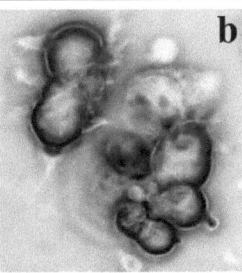

Figure 20 Microscopic images of HEK cells. (a) unsheared control culture (b) sheared culture after applying 60 passes through a capillary with an inner diameter equal to 0.51 mm and a flow rate of 5.94 mL min^{-1} resulting in a hydrodynamic stress of 9.22 Pa.

Similar observations of smaller and rougher, obviously apoptotic cells, were presented by Al-Rubeai et al. (1995) using hybridoma cells cultivated in an agitated vessel. This observation is further supported by the measurements of the fluorescence from AnnV and PI binding. An example of the measured AnnV and PI fluorescence for the unsheared and sheared cells is presented in *Figure 21*. As can be seen higher values of the hydrodynamic stress lead to a significant increase of the AnnV fluorescence (with respect to the unsheared control sample). The peak maximum of the fluorescence signal is clearly shifted to higher values. On the other hand, no change of the PI fluorescence (right column) is observed and supports the previous statement that the most probable mechanism for HEK cell death, under the applied conditions, is apoptosis.

By applying capillaries with the same inner diameter but different lengths the effect of the exposure time on the cell damage is investigated. HEK cells are exposed to extensional flow at 0.49 Pa, which is below the threshold, and at 9.22 Pa, which is above the limit. After 60 passes through the capillary the cell samples are analyzed by FACS as described previously. As shown in *Figure 22*, no influence of the capillary length on the extent of cell death is observable, which is in good agreement with the results published by Zhang et al. (1993) and Aloi and Cherry (1994). This supports the CFD analysis, where the largest values of the hydrodynamic stress are observed at the entrance or exit of the capillary.

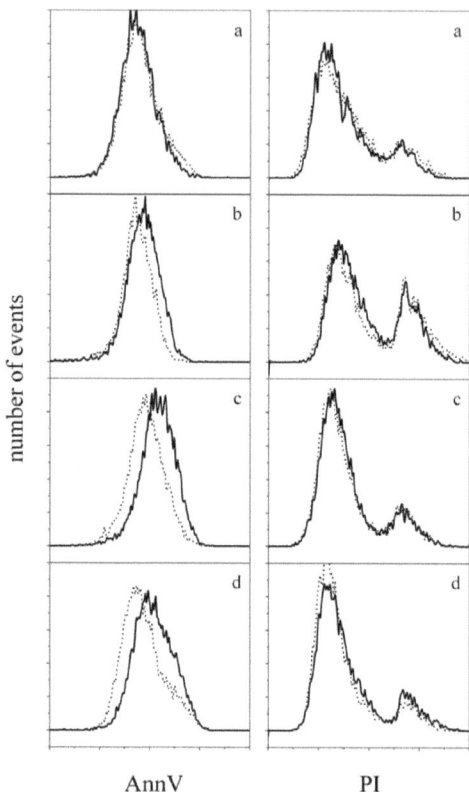

*Figure 21 FACS analysis of HEK cells: left column: AnnV fluorescence, right column: PI fluorescence. (---) unsheared control sample; (—) sheared sample with an applied shear stress equal to (**a**) 0.49 Pa (**b**) 1.09 Pa (**c**) 3.05 Pa (**d**) 9.22 Pa.*

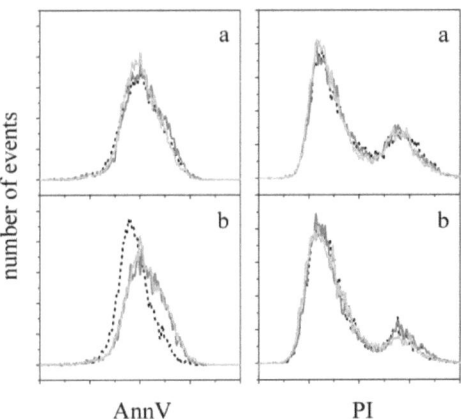

Figure 22 FACS analysis of HEK cells. Influence of the capillary length on the extent of cell death. (a) at 0.49 Pa and (b) at 9.22 Pa. (---) unsheared control sample, (—) sheared at 1x capillary length, (—) sheared at 3x capillary length.

To test whether similar phenomena would be applicable also for CHO cells, equivalent experiments for this cell line are performed. As can be seen from *Figure 23* also for this cell line a slight increase of the AnnV fluorescence is measured for sheared cells with respect to unsheared cells while no change in the PI fluorescence, for both sheared and unsheared samples, over the whole range of hydrodynamic stress is observed. As can be seen for this particular cell line the threshold value of increased AnnV fluorescence is around 3.05 Pa (FSC vs. SSC not shown). To further support this observation, *Figure 24* shows images of CHO cells, unsheared and sheared at an average hydrodynamic stress τ_L^{ave} equal to 9.22 Pa. Intensive cell blebbing, the loss of halo, shrinkage in cell size and increased cell granularity can be noticed. The obtained pictures of the sheared cells are again similar to the images of cells obtained from Staurosporine-induced apoptosis (see section 2.2.5).

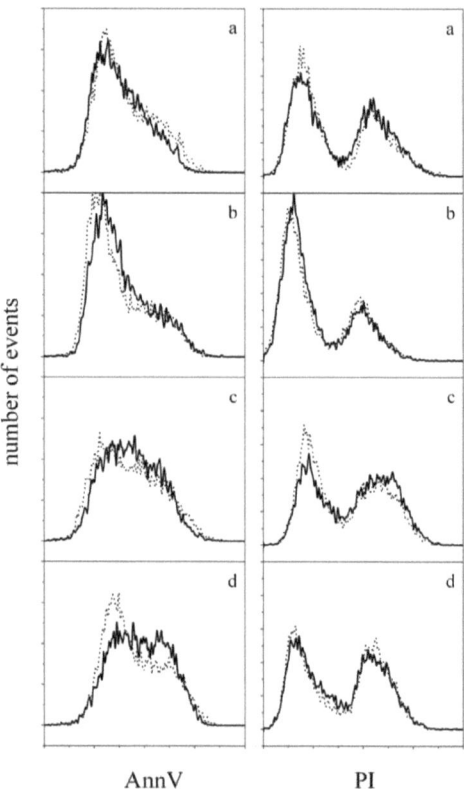

Figure 23 FACS analysis of CHO cells: left column: AnnV fluorescence, right column: PI fluorescence. (---) unsheared control sample; (—) sheared sample with an applied shear stress equal to (a) 0.49 Pa (b) 1.09 Pa (c) 3.05 Pa (d) 9.22 Pa.

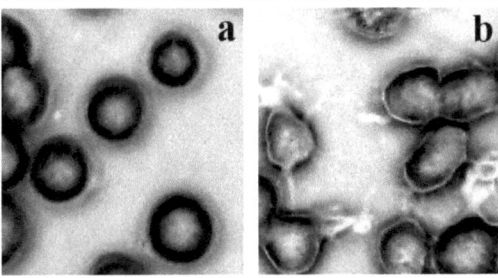

Figure 24 Images of CHO cells. (a) unsheared control culture; (b) sheared culture after applying 60 passes through a capillary with an inner diameter equal to 0.51 mm and a flow rate of 5.94 mL min^{-1} resulting in a hydrodynamic stress of 9.22 Pa.

It can be concluded that for both cell lines the most probable mechanism controlling cell death, when exposed to low levels of hydrodynamic stress originating from the mean velocity gradient with short duration, is apoptosis. Similar to the simple shear flow also for the extensional flow, a difference in the average hydrodynamic stress threshold values for apoptosis for HEK and CHO cells, equal to 1.09 Pa and 3.05 Pa, respectively, is found. These results accompany the results published by Aloi and Cherry (1994), who detected an increased intracellular calcium concentration, which is known to be a preceding event during apoptosis (Trump and Berezesky, 1995), as response of Sf-9 cells to extensional flow (covering both laminar and turbulent conditions). As most of the previous research showed only necrotic cell death in extensional flow fields (McQueen et al., 1987; Zhang et al., 1993), this death mechanism is studied here as well by using a capillary with a diameter of 0.20 mm and employing higher flow rates. When applying a hydrodynamic stress above a certain threshold value both cell lines suffer severe necrosis as measured by an increased DNA release, $\Theta_{s/ns}$, (*Figure 25*), as well as and by an increase in AnnV and PI fluorescence (*Figure 26*).

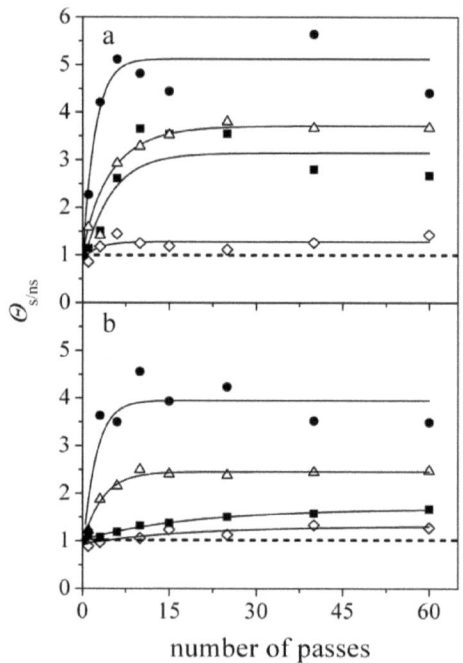

Figure 25 Example of DNA release kinetics upon shear exposure by applying extensional flow. (a) HEK cells; (b) CHO cells. (—) exponential fit (●) 2588 Pa (△) 1719 Pa (■) 1007 Pa (◇) 237 Pa. Detailed information about the operating conditions are shown in Table 2.

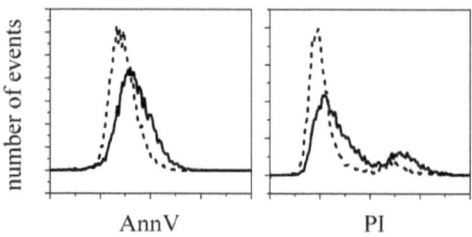

Figure 26 FACS analysis of HEK cells, PI and AnnV fluorescence. (---) unsheared control sample; (—) sheared sample after 60 passes through the sudden contraction applying a hydrodynamic stress equal to 1007 Pa.

Chapter 3 - Response of Mammalian Cells to Environmental Stress

In general it is required to apply between 3 and 15 passes to reach a steady state. Connecting this observation with the values of the averaged hydrodynamic stresses as presented in *Table 2*, it can be seen that significant cell damage and therefore an increase of $\Theta_{s/ns}$, occurs when τ_{is}^{ave} becomes dominant, with a ratio of cell size to the Kolmogorov length scale around 5. Comparable ratios of the cell size to Kolmogorov length scale for cell necrosis were found by other authors (McQueen et al., 1987; Zhang et al., 1993). To summarize the observed trend, a comparison of the steady state values of $\Theta_{s/ns}$ as a function of the applied hydrodynamic stress is presented in *Figure 27*. It can be seen that for both cell lines the threshold value of the hydrodynamic stress is equal to 500 ± 100 Pa. The uncertainty interval is due to the cell size distribution covering a range from 7 to 15 µm. As the relevant data in the literature do not use CFD to characterize the flow the comparison with this threshold value is not straightforward. For that purpose corresponding values of the wall shear stress, τ_{wall} (see *Table 2*), which for the threshold is around 300 Pa, are used. This value is in agreement with τ_{wall} values around 180 and 260 Pa as obtained by McQueen et al. (1987) and Zhang et al. (1993) using hybridoma and myeloma cells. No effect of the capillary lengths on the extent of cell death was also observed by Zhang et al. (1993) and Aloi and Cherry (1994), but contrasted with the results by McQueen et al. (1987).

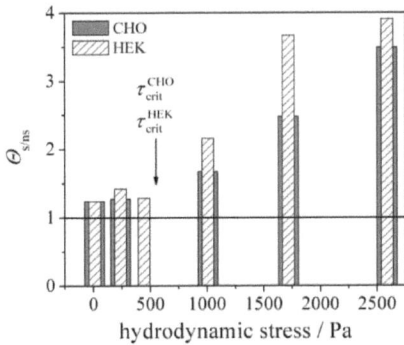

Figure 27 Average $\Theta_{s/ns}$ from CHO and HEK cells upon shear exposure after applying 60 passes through the sudden contraction (exp. error within 17 %) as a function of average hydrodynamic stress (Table 2).

55

Approximating the cell bursting in the experiments with the disintegration of drops or bubbles in the inertial subrange of turbulence, when the hydrodynamic stress exceeds the confinement stress τ_c $(=6\sigma_m/d_{cell})$ (Lasheras et al., 2002), it is possible to estimate the value of the cell membrane tension, σ_m. Using the value of the critical hydrodynamic stress as determined above and a minimum and maximum cell size d_{cell} equal to 7 and 15 µm the cell membrane tension σ_m for these two cell lines was equal to 1.17 ± 0.65 mN m^{-1}. This value is in good agreement with values measured by Zhang et al. (1992), where the bursting tension of individual hybridoma cells using a micromanipulation technique equal to 1.8 ± 0.5 mN m^{-1} was found.

3.2.4 Cell Response to Extensional Flow (Long Term)

Since, the critical threshold values presented above are determined in short term experiments, where the cells are sheared only for a short time of 1 - 4 hours, it is of practical interest to investigate the response of the cells over long term exposure which should reflect more closely the hydrodynamic environment a cell will encounter during culturing inside a sparged stirred tank reactor (Aloi and Cherry, 1996). In this type of experiment (*Figure 8*) the exposure to high shear stresses inside the capillary will represent the peak values close to the impeller, whereas the reservoir is mimicking the bulk fluid featuring the lowest shear stresses. In consequence, the cells will encounter two distinct hydrodynamic regions in an alternating fashion. During the extended shear experiments, as described in section 2.6, the culture is pumped through the capillary by a syringe pump over a period of 1500 passes. The circulation time of a suspended cell within a STR, operated under turbulent conditions, can be approximated to be equal to a fifth of the reactor mixing time t_M (Campolo et al., 2003; Nienow, 1997; Patwardhan and Joshi, 1999). Taking a typical t_M for a STR of around 60 seconds, it can be concluded that the experimental set-up is able to mimic the *in vitro* environment within a stirred tank reactor over a period of 5 hours. Throughout the experiment, the culture fluid was oxygenated by surface aeration inside the reservoir and oxygen limitation could be eliminated. Additionally, the temperature was kept at 37° C. At the end of each experiment the pH of the supernatant was determined to exclude the effect of sub-optimal culture environment. Moreover, one control culture was always used to exclude any side-effects of a sub-optimal culture environment.

In contrary to the short term experiments, multiple exposures over a long time lead to a lower threshold as presented in *Figure 28*. Applying shear stress values lower than 1 Pa, the cell viability is constant over a long time. The forces acting on the cell are not high enough to cause any damage. In contrast, when applying a hydrodynamic stress of 10 Pa, macroscopic damage to the cellular membrane can be observed leading to a decreased viability to 30 % after 1500 passes. Moreover, exposure to even higher stresses around 170 Pa leads to a final viability of 20 % after the same number of passes. Taking the previous experimental results into account, where the same cell line suffered apoptosis after short term exposure (60 passes) to a hydrodynamic stress of around 2 Pa, this would indicate that the culture exhibits post-apoptotic necrosis. Another contribution could be the cumulative effect of multiple and continuous exposure to sub-lethal shear stress levels resulting eventually in necrosis.

The decrease in cell viability is confirmed by the measurement of released DNA into the supernatant (*Figure 29*). Clearly, the critical value can be found in the range of 1 to 10 Pa. Nevertheless, the exact threshold for a particular cell line is influenced by many parameters like medium composition or culture history and has to be always determined independently.

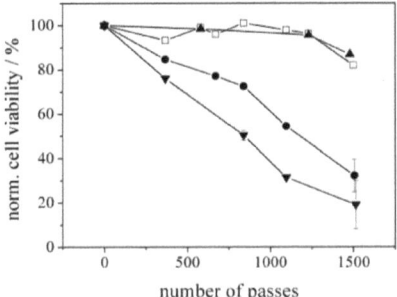

Figure 28 Cell viability of CHO cells upon shear exposure after applying 1500 passes through the sudden contraction as a function of the hydrodynamic stress. (□) 0.03 Pa (▲) 1.0 Pa (●) 10 Pa (▼) 170 Pa.

Chapter 3 - Response of Mammalian Cells to Environmental Stress

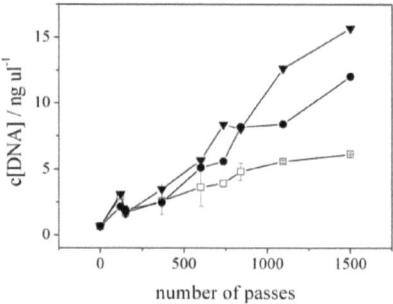

Figure 29 DNA release of CHO cells upon shear exposure after applying 1500 passes through the sudden contraction as a function of the hydrodynamic stress. (□) 0.03 Pa (●) 10 Pa (▼) 170 Pa.

By the addition of Pluronic F-68 to the culture medium the detrimental effect of the shear stress on cell viability could not be significantly lowered (data not shown). This is probably due to the fact that Pluronic is known to protect cells only under sparging conditions where the cells are damaged as a result of the contact with rising and bursting bubbles (Michaels and Papoutsakis, 1991).

From the practical point of view, it can be summarized that a particular bioreactor should not exhibit regions featuring peak shear stresses, τ_{crit}, above 1 Pa, to avoid cell damage due to hydrodynamic forces.

3.3 Conclusions

It was shown that protein-free growing CHO and HEK cells respond differently when exposed to steady simple shear or extensional flow patterns. Exceeding a critical threshold value, the cultures will either enter the apoptotic pathway or dye by necrosis. In simple shear flow, low values of hydrodynamic stress (τ_{crit} around 2 Pa), lead to a permanent deformation of the cell and cause severe cell necrosis. In contrast, when cells are exposed to extensional flow with hydrodynamic stresses of the same amplitude, operated under laminar conditions, a very short exposure to the hydrodynamic stress leads to the induction of the apoptotic death pathway. In extensional flow necrosis was either observed when the cell size is larger than the Kolmogorov length scale resulting in significantly larger values of the critical hydrodynamic stress with respect to the simple shear case ($\tau_{crit} > 500$ Pa) or when the exposure time was significantly increased (1 Pa $<\tau_{crit}<$ 10 Pa). Based on the measurements presented in this study, the CHO cell line proved to be more robust against the induction of apoptosis but suffers necrosis under slightly lower hydrodynamic stress compared to the HEK cell line.

The maximum shear levels determined in this study should be kept in mind and avoided during pharmaceutical production to enhance culture longevity and to increase recombinant protein productivity.

Chapter 4
Development of a Novel Taylor-Couette Bioreactor

4.1 Introduction

4.1.1 Bioreactor Design Considerations

The expansion of all types of cells requires suitable bioreactors to provide an optimal *in vitro* environment, mimicking *in vivo* conditions. Ideally, the bioreactor should provide and maintain the highest degree of environmental uniformity, avoiding concentration gradients of nutrients, toxic metabolic byproducts, and oxygen, without featuring critical levels of shear stress (*Figure 30*). Since all these parameters are interrelated, compromises have often to be made, leading necessarily to sub-optimal performance.

One of the most critical design parameter is the oxygen mass transfer rate. The major aim of aeration is to ensure the oxygen supply to the cultured cells, needed for their energy metabolism in aerobic pathways (e.g. the oxidative phosphorylation in the glucose metabolism). The typical oxygen demand of cultured mammalian cells is reported to be equal to 2.6×10^{-13} mol O_2 cell^{-1} h^{-1} (Ducommun et al., 2000). For optimal cell growth a dissolved oxygen concentration (DO) of 20 - 80 % air saturation (AS) should be adjusted (Ozturk and Palsson, 1990). A lower critical DO was found at 0.5 % AS (Miller et al., 1987).

Chapter 4 - Development of a Novel Taylor-Couette Bioreactor

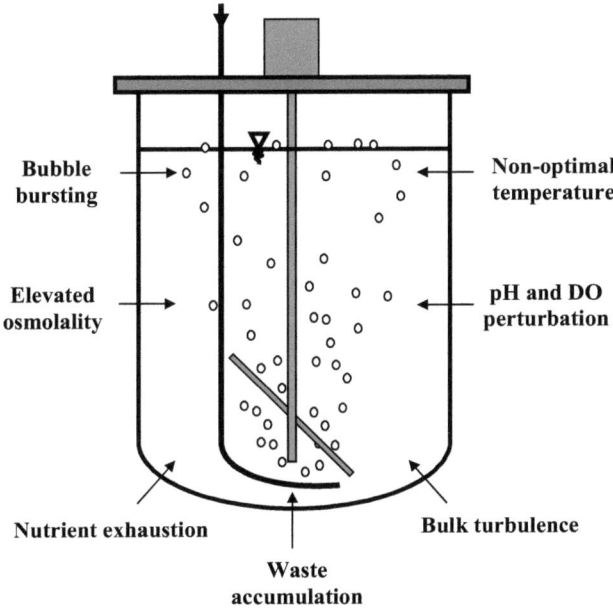

Figure 30 Cell debilitative forces within the bioreactor environment.

Moreover, in any cultivation vessel mixing is crucial, determining heat and mass transfer, culture performance, and product quality. In general, mixing pursues three major goals: 1) maintaining particles, e.g. cells, in suspension 2) the homogenous distribution of nutrients and metabolic waste products by minimizing gradients and 3) dispersing the medium components on the molecular level (micro-mixing). The mixing time should be significantly smaller than the time scales of cell growth and substrate consumption, to avoid transport limitations. However, for the cultivation of mammalian cells typical rates are in the scale of hours and are therefore rather limited by kinetics than by transport. Nevertheless, liquid homogeneity has to be ensured to expose the cultured cells to a consistent microenvironment, which will guarantee highest product quality and reproducibility of the process (Andersen and Goochee, 1995; Borys et al., 1993; Cruz et al., 2000; deZengotita et al., 2002; Lara et al., 2006; Miller et al., 1987).

4.1.2 State of the Art in Biopharmaceutical Development and Production

In most biopharmaceutical applications the stirred tank reactor (STR) is the system of choice and is used as major workhorse for the cultivation of mammalian cells and the production of recombinant proteins. Furthermore, cell-based technologies using mammalian tissues and stem cells for therapeutic applications are gaining attention and become a focal point of interest within the pharmaceutical industry.

Although well established in industrial practice, the STR exhibits several drawbacks such as mixing pathologies (Alvarez et al., 2005; Arratia et al., 2004), foaming problems (Nehring et al., 2004), difficult scalability under laminar conditions (Alvarez et al., 2005; Arratia et al., 2004), a broad distribution of the energy dissipation rate (Alexopoulos et al., 2002; Bouyer et al., 2005; Cutter, 1966; Wu and Patterson, 1989; Zhou and Kresta, 1998), and the potential of damaging shear sensitive cells, tissues and cell aggregates in the zone of highest hydrodynamic stresses in the vicinity of the impeller and through bubble aeration. Besides bulk hydromechanical forces (for instance high shear gradients and disruptive eddies) in the vicinity of the impeller, shear forces originating from bubble breakup and in draining films during foam formation being, at least one order of magnitude greater those in the bulk (Marks, 2003), cause sub-lethal cell response (Dunlop et al., 1994; Joosten and Shuler, 2003; Keane et al., 2003; Senger and Karim, 2003) or in the severest case cell death (Al-Rubeai et al., 1995; Cherry and Hulle, 1992; Chisti, 2001; Garciabriones and Chalmers, 1994; McQueen et al., 1987; Tramper et al., 1986).

To overcome the mentioned problems several bioreactors for the production of biopharmaceutical products have been proposed and implemented, including various stirred tank modifications, airlift, hollow-fiber, packed-bed, fluidized-bed, and most recently different disposable devices, like the Wave bioreactor. A detailed summary of these configurations can be found elsewhere (Warnock and Al-Rubeai, 2006). Possible solutions to replace bubble aeration are membrane-based air supply systems employing tubes or flat membranes inside the reactor or external oxygenation modules. Here, silicone, polypropylene (PP), and polytetrafluoroethylene (PTFE) tubing and membranes have turned out to be the most appropriate materials because of their high gas permeability and their mechanical, thermal, and chemical stability (*Table* 3).

Material	Cell system	D_{wall} / mm	A/V ratio	$k_L a$ / h^{-1}	Ref.
PP hollow fiber	Plant cells	-	24.5	2.64 - 3.75	(12)
PP hollow fiber	Mouse fibroblast	0.8	8.2 - 32.7	1.93 - 7.74	(2)
PP membrane	-	-	33.8	7.92	(10)
PP membrane	Hybridoma	-	5.9 - 11.8	0.34 - 0.8	(8)
PTFE tubing	Hybridoma	0.5	5.0	6.1 - 7.0*	(5)
PTFE tubing	CHO	0.5	12.2	0.63 - 1.58	(3)
Silicone, PP tubing	CHO, hybridoma	-	16.4 - 28	1.6 - 3.8	(14)
Silicone tubing	Plant cells	0.4	13.91	0.44	(11)
Silicone tubing	Plant cells	0.4	14.3	-	(17)
Silicone tubing	-	1.195	66.1	5.0	(4)
Silicone tubing	MDCK	0.8	25.1	1.3 - 1.6	(6)
Silicone tubing	Hybridoma	0.7	69.1	-	(7)
Silicone tubing	Hybridoma	0.35	31.6	3.0	(13)
Silicone tubing	Mouse LS cells	0.25	4.8	1.18 – 11.8	(9)
Silicone tubing	FS-4	0.48	2.7	~0.5	(15)
Wave bioreactor	NS0, HEK, Sf9	-	-	0.5 - 4.0	(16)
Surface aeration	Murine fibroblast	-	0.2	0.2 - 0.8	(1)
Ceramic microsparger	MDCK	-	-	30.6 - 50.4	(6)
Stainless steel sparger	MDCK	-	-	9.4 - 10.4	(6)

Table 3 Comparison of sparging and bubble-free oxygenation approaches. (1) (Curran and Black, 2004) (2) (Lehmann et al., 1987) (3) (Ducommun et al., 2000) (4) (Qi et al., 2003) (5) (Schneider et al., 1995) (6) (Nehring et al., 2004) (7) (Zhang et al., 1993) (8) (Beeton et al., 1991) (9) (Kilburn and Morley, 1972) (10) (Millward et al., 1996) (11) (Luttman et al., 1994) (12) (Bohme et al., 1997) (13) (Persson and Emborg, 1992) (14) (Fenge et al., 1993) (15) (Fleischaker and Sinskey, 1981) (16) (Singh, 1999) (17) (Hvoslef-Eide et al., 2005).

* $k_L a$ values from Nehring et al. (2004) calculated from exp. data of Schneider et al. (2004).

Despite, the efforts made, there is still a need for further reactor development to find the best compromise between maximized liquid homogeneity and mass transfer while exposing the cultured cells and tissues to a minimum level of shear stress. Addressing these requirements, the concept of the Taylor-Couette (TC) like bioreactor with a bubble-free aeration system was first introduced and successful implemented for the cultivation of red beet suspension cells (Janes et al., 1987). Later on, the rotating-wall vessel (RWV) was developed and patented (Anderson and Schwarz, 1998; Schwarz et al., 1988). A similar concept was introduced with the high-aspect rotating-wall vessel (HARV) reactor, featuring no inner cylinder (Cowger et al., 1999; Cowger et al., 1997). For Sf-9 cells it was shown that in shaker flasks the necrotic cell formation rate constant was three times higher than in the HARV culture. Additionally, the apoptotic cell formation rate constant turned out to be almost 4 times lower in the HARV culture, which was attributed to the reduction of hydrodynamic stress exerted to the cell population. These observations were accompanied by reduced glucose utilization, lower yields in lactate and ammonia, and enhanced culture longevity. Baby hamster kidney (BHK) cells were cultivated on microcarriers under low (0.05 Pa) and high shear (0.09 Pa) conditions within a rotating wall vessel (Goodwin et al., 1993). It was found that under high shear the formation of 3-dimensional aggregates was suppressed, the cells failed to proliferate, and the release of intracellular components was increased. The growth and productivity of two insect cell lines was investigated in shaker flasks and the HARV (Saarinen and Murhammer, 2000). It was concluded that the lower shear, present in the HARV, may be the beneficial factor for increased protein production. Published data indicate that the application of rotating wall vessels and its gentle, homogenous microenvironment may contribute to increased protein production by allowing the cells to spend their metabolic energy on growth and production rather than for repair mechanisms required after stress exposure.

Nevertheless, rotating wall vessels and Taylor-Couette devices are suffering from several drawbacks. For instance, it was shown that the RWV, employing a concentric inner cylinder composed of a gas permeable membrane, where the mass transport is controlled by molecular diffusion, develops a very heterogeneous oxygen distribution within the working volume, which decreases sharply towards the outer cylinder wall. This is the case even for cultivations containing cell aggregates or cell supporting microcarriers when operated at low agitation rates or small particle sizes (Kwon et al., 2008). Additionally, all systems operating with a rotating outer cylinder will be difficult to operate under continuous or perfusion mode, due to construction problems. Considering the shear rate distribution, TC devices as well as ST reactors feature in general a broad profile leading to a heterogeneous microenvironment (Soos et al., 2007). Moreover, limitations of the TC unit are emerging when operated under laminar conditions, where it was shown that intra-vortex mixing is dominated by molecular diffusion, leading to significant heterogeneities (Desmet et al.,

1996). Further limitations were identified for two phase systems containing particles, where the dispersed phase accumulated in regions of low velocities, that is the vortex cores (Resende et al., 2001). To prevent such a segregation process it is possible to use the eccentrical assembly of the inner cylinder (Cole, 1967; Tennakoon and Andereck, 1993; Vohr, 1968), the rotation of both cylinders at different velocities (Lopez and Marques, 2002) and the implementation of a non-circular cross-section of the inner cylinder (Snyder, 1968; Soos et al., 2007).

4.1.3 Design and Construction of the LTC Reactor

In this work the concept of the lobed Taylor-Couette reactor (Soos et al., 2007) is extended to biotechnological applications so as to overcome the limitations mentioned above present for TC and ST units. For a better understanding, a brief introduction into the concept of the Taylor-Couette system shall be given in the following.

The fluid motion in the annular space between two coaxial cylinders has been already described in the 19th century and found its first application as viscometer in 1890 (Couette, 1890; Taylor, 1923). Assuming a static outer cylinder and a slowly rotating inner one, the developed flow becomes steady and purely azimuthal and the velocity gradient in radial direction, also called shear rate, γ, is equal to the derivative of the angular velocity with respect to the radial direction. Any disturbance by the velocity is damped by the fluid viscosity and the flow is called circular Couette flow (*Figure 31a*). It is worth noting that the same type of flow develops when the outer cylinder is rotating and the inner one is at rest. Accelerating the rotation speed of the inner cylinder, conditions are reached where the viscous forces, which are proportional to the velocity gradient, are not any more able to compensate the centrifugal force acting on the fluid elements, which is proportional to the square of the velocity, leading to the development of a secondary flow. This flow pattern characterized by the presence of axisymmetric vortices, known as Taylor vortex flow (Taylor, 1923), is superimposed on the Couette flow as illustrated in *Figure 31b,c*.

Increasing the rotation speed further a second transition can be observed, where the vortices start to oscillate generating an azimuthal wave, traveling around the inner cylinder known as wavy vortex flow. The number of azimuthal waves and the individual vortex height depend upon the cylinder length, the rotation speed and d/r_i (Di Prima and Swinney, 1979). Eventually, when the rotation speed is increased even further, the flow becomes

Chapter 4 - Development of a Novel Taylor-Couette Bioreactor

gradually turbulent and is called turbulent vortex flow. A more detailed description of the different pattern and transitions in coaxial cylinders flows can be found elsewhere (Andereck et al., 1986; Kataoka, 1986). As for other flow devices such as pipes or stirred tanks, the observed flow pattern can be related to the dimensionless Reynolds number (*Re*), relating the inertial forces relative to the viscous forces, defined as:

$$Re = \frac{r_i \varpi d}{\nu} \qquad (24)$$

where r_i represents the radius of the inner cylinder, ϖ the annular velocity, d the gap width, and ν the kinematic viscosity.

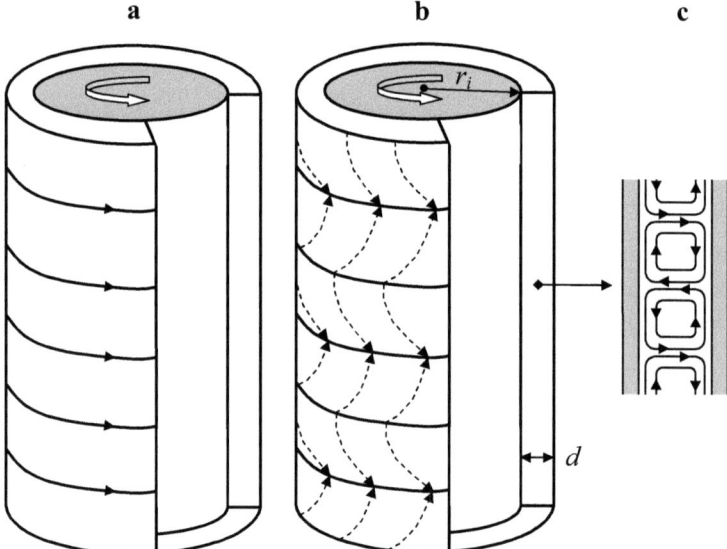

*Figure 31: Flow pattern between to concentric cylinders. (**a**) circular Couette flow; (**b**) Taylor vortex flow; (**c**) vortex cells developing in the gap during Taylor vortex flow.*

To relate centrifugal forces relative to the viscous forces, another dimensionless group called Taylor number (*Ta*), is used. This is defined as following:

$$Ta = \frac{r_i \varpi^2 d^3}{\nu^2} \qquad (25)$$

66

As pointed out in section 4.1.2, the classical TC suffers from several drawbacks. Addressing these problems, a Taylor-Couette unit with three exchangeable inner cylinder modules of different cross-sectional shape was designed, characterized, and patented from our laboratory (Morbidelli et al., 2005; Soos et al., 2007). The different devices are used to evaluate the influence of the cross-sectional shape of the inner cylinder on the flow characteristics within the annular gap. The respective cross-sections are given in *Figure 32*. The left one corresponds to the classical TC with a constant gap width equal to 8 mm. The other two represent two lobed configurations with a non-constant gap width (8-12 mm and 8-20 mm). Accordingly, the three devices are named TC, LTC 8-12 and LTC 8-20 with d_{max}/d_{min} equal to 1.0, 1.5 and 2.5.

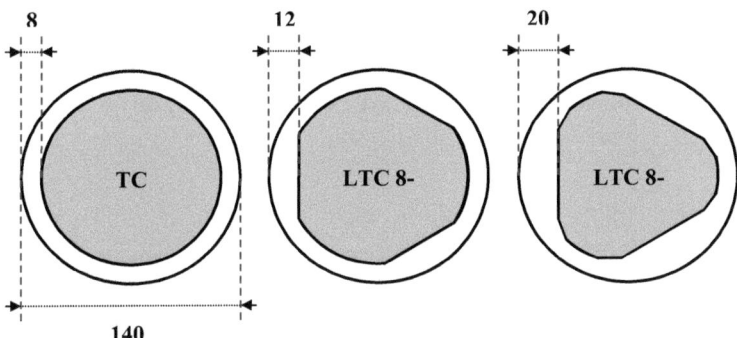

Figure 32 Schematic view of the Taylor-Couette reactors with different cross-sections - all dimensions in mm.

To characterize the hydrodynamic environment inside the three units, several experiments and simulations have been carried out and shall be discussed in the following.

By the use of a rheoscopic fluid, the flow field within the three reactors is visualized. To emphasize the improvement made through the implementation of a non-circular cross-section of the inner cylinder with a large d_{max}/d_{min}, the flow profile of the LTC is compared with a common TC (d_{max}/d_{min} = 1) and a LTC with a smaller d_{max}/d_{min} (see *Figure 32*). In *Figure 33* the flow regimes, occurring at different rotational speeds (rpm) in the three reactor configurations are given.

The flow pattern is commonly related to Re/Re_c (Kataoka, 1986), which for instance has to be greater than 20 for r_i/r_o around 0.85 (Lueptow et al., 1992) for turbulent Taylor vortex flow. For a radius ratio equal to 0.68 the turbulent Taylor vortices appear at Re/Re_c around 35 (Kataoka, 1986).

At the lowest rotation speed during these experiments, Re/Re_c in the LTC 8-20 is equal to 63, meaning the device is operating under turbulent vortex flow. To obtain the same Re/Re_c in the LTC 8-12 and the TC, it is necessary to increase the rotation speed to 90 and 150 rpm, respectively. By comparing the respective pictures in *Figure 33*, it becomes obvious that under these conditions the vortex instabilities in the LTC 8-20 are more pronounced. For the TC and the LTC 8-12, Taylor-vortices are clearly visible and dominating the flow pattern, leading to a reduction of intra and inter vortex mixing (Desmet et al., 1996; Snyder, 1968). Overall, the application of a non-constant gap width along the radial direction perpendicular to the rotation axis leads to additional instabilities superimposed to the main flow structure (Taylor vortices) in the TC device.

Chapter 4 - Development of a Novel Taylor-Couette Bioreactor

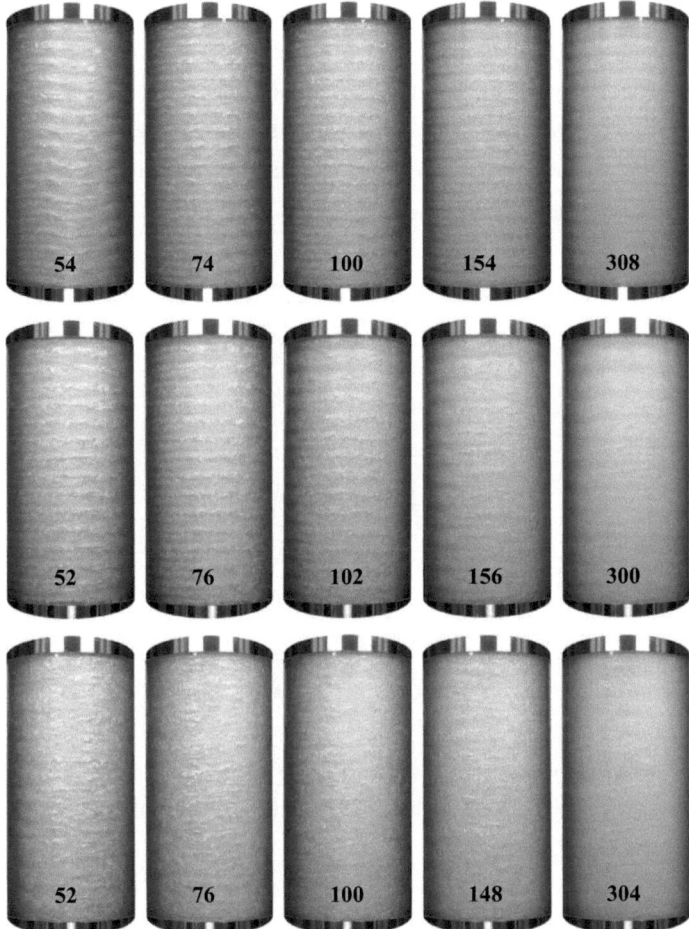

Figure 33 Flow regimes, occurring at different rotational speeds (rpm) in the three reactor configurations, visualized by the use of a rheoscopic fluid. Numbers in bold represent the rotation speed in rpm. Upper row: TC 8-8, middle row: LTC 8-12, lower row: LTC 8-20.

As mixing within a bioreactor is a critical process parameter it is characterized by determining the mixing times in all three units as described in 2.5. In *Figure 34* the normalized pH response curves after disturbance, measured at the same volume average energy dissipation rate ε (=0.013 m^2 s^{-3}), are shown. All three response curves show a certain delay after the base injection, as the tracer has to travel through the annular gap to the other end of the unit, where the pH probe is installed. Clearly, the stable vortices within the TC unit (see *Figure 33*) hinder the transport and increase the time needed for homogenization. By increasing d_{max}/d_{min} these vortices are destabilized and the mixing is enhanced, as seen for the LTC 8-12 and LTC 8-20.

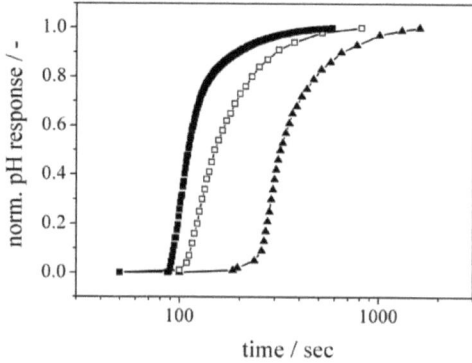

Figure 34 Normalized pH response curves after disturbance, measured at the same volume average energy dissipation rate ε (=0.013 m^2 s^{-3}). (▲) TC (□) LTC 8-12 (■) LTC 8-20.

Another quantity that has to be determined is the local hydrodynamic stress inside the culture unit, which is known to be deleterious for the suspended cells (3.2.4). For a detailed description of the occurring shear stresses see section 3.1.2. As the flow in all devices is three dimensional, the maximum positive eigenvalues of the rate of strain tensor are evaluated numerically applying the Householder reduction and the QL method (Press et al., 1992) implemented through a User Defined Function (FLUENT 6.2. UDF Manual 2005). To evaluate the energy dissipation rate in a particular system it is common to use either a torque measurement, or an empirical correlation for the Power number (Langheinrich et al., 2002; Nienow, 1997), or Computational Fluid Dynamics (CFD) simulations. As it was shown previously, the LTC 8-20 shows a substantially narrowed distribution, meaning a more homogenous environment, with respect to the other two devices (Soos et al., 2007).

4.1.4 Design of the LTC Bioreactor

In the studies made above, the LTC 8-20 showed superior performance over the TC and the LTC 8-12 units, regarding the exhibited flow structure, the mixing time and the distribution of the energy dissipation rate (Soos et al., 2007) and can be therefore regarded as improved design. Based on these results, a lobed Taylor-Couette reactor is constructed featuring a d_{max}/d_{min} equal to 2.5. The LTC reactor has a total working volume of 3.0 L and consists of two concentric cylinders as illustrated in *Figure 35*, and *Figure 36*, adapted from the LTC 8-20.

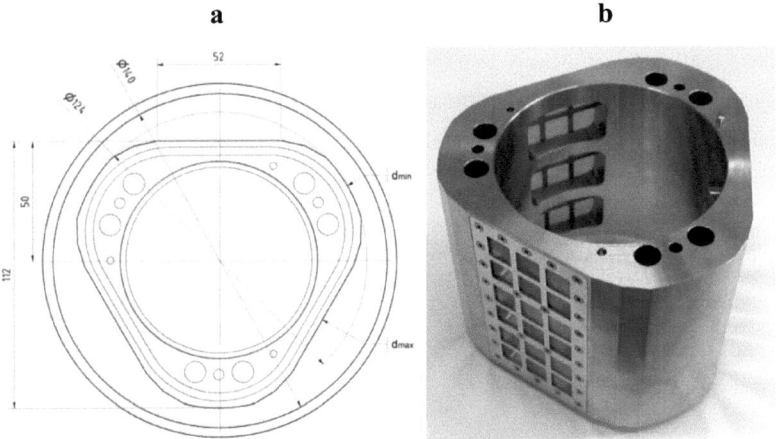

Figure 35 (a) cross-section of the LTC reactor - all dimensions in mm; (b) one part of the inner cylinder with installed membrane sheets.

Rotation of the inner cylinder is provided via a 3-phase step motor (Berger Lahr, Germany). The reactor has no head space and the ends of the reactor are conically shaped (*Figure 36*). Oxygen is delivered via a steel-mesh reinforced silicone membrane (Franatech GmbH, Germany) with a thickness of 150 μm ± 5 μm, mechanically fixed on the perforated inner cylinder (*Figure 35* and *Figure 36*). The specific oxygen permeability of the membrane as provided by the manufacturer is equal to 0.26 mol O_2 m^{-2} h^{-1} bar^{-1}. To deliver enough oxygen to the cultivation media the air pressure of the inner cylinder is regulated and stepwise increased on demand of the growing culture. The culture temperature of 37° C is controlled using a tube, wrapped around the outer cylinder connected to the Labfors® temperature control unit. The bioreactor can be operated in either vertical or horizontal mode. To measure and control all relevant quantities during cultivation at each end of the vessel a pH electrode (Lazar Research Labs, USA), a DO sensor (Mettler-Toledo, Switzerland) and a temperature probe (Condustrie-Met, Switzerland) are mounted (*Figure 36*). In addition bottles for sampling, re-feeding, base (0.2 mol NaOH) and acid (0.2 mol HCl) addition, inoculum and feed/harvest are connected via tubing to the reactor ports. On the top of the reactor a liquid level indicator is mounted and the gas-cooler attached. A picture of the final set-up can be seen in *Figure 37*.

To calculate the membrane area required for batch cultivation of mammalian cells a maximal achievable cell density equal to 5×10^6 cells mL^{-1} (Nienow, 2006) is assumed. Using a typical oxygen uptake rate of 2.6×10^{-13} mol O$_2$ cell^{-1} h^{-1} (Ducommun et al., 2000) and the specific oxygen permeability of the membrane stated by the manufacturer, a necessary membrane area of 147 cm^2 is calculated. The reactor features a total membrane area of 297 cm^2 and offers therefore a safety buffer of a factor two. In addition, the aeration capacity could be increased by feeding oxygen enriched air instead of pure air.

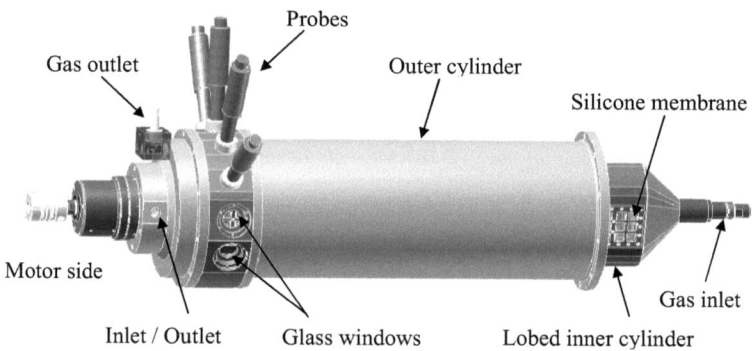

Figure 36 Schematic set-up of the LTC reactor.

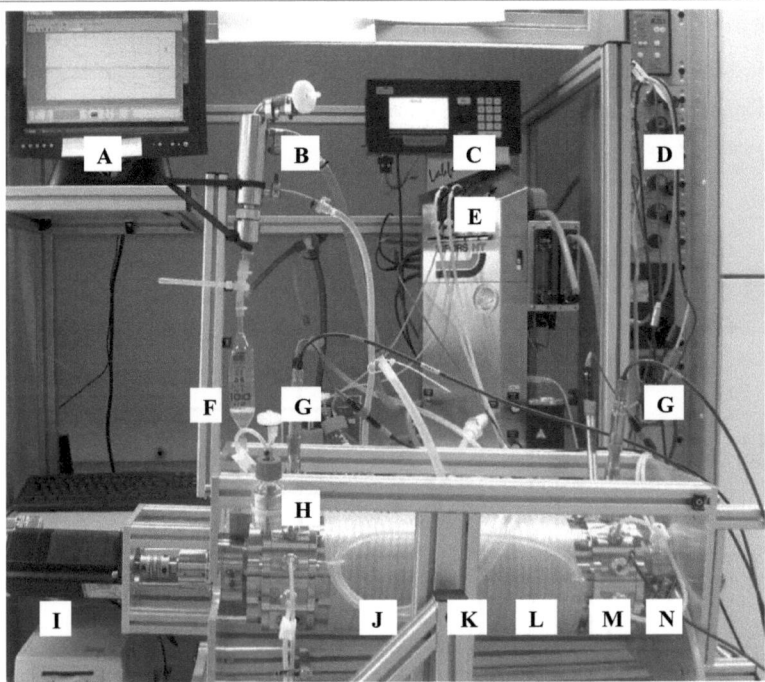

*Figure 37 Set-up of the LTC. **A** Monitoring; **B** Gas cooler; **C** Control unit; **D** Gas pressure valve; **E** Pumps; **F** Liquid level indicator; **G** Probes (pH, DO, T); **H** Sampling device; **I** Motor; **J** Rotary axis; **K** Temperature control; **L** Observation window; **M** Feed and harvest lines; **N** Gas inlet.*

4.1.5 Stirred Tank Reactor

A commercial stirred tank benchtop reactor (Labfors® III, Infors HT, Switzerland) with a total volume equal to 3.6 L (2.7 L working volume) is used as benchmark reactor (*Figure 38* and *Figure 39*). The double-jacketed, round bottom vessel is equipped with a pitched, elephant-ear three-bladed impeller, magnetic coupled with the motor. The gas flow rate for aerating the STR is controlled by a thermal mass flow valve, Red-y, with a complete mass flow range between 0.06 - 2.0 L min^{-1} (corresponding to 0.02 - 0.74 volume per volume and minute (vvm)). The agitation rates are adjusted to values between 40 and 300 rpm. In the lid of the vessel a pH electrode (Lazar Research Labs, USA), a DO sensor

(Mettler-Toledo, Switzerland) and a temperature probe (Condustrie-Met, Switzerland) are mounted. Additionally, bottles for sampling, base (0.2 mol NaOH) and acid (0.2 mol HCl) addition, inoculum and feed/harvest are connected via tubing to the reactor ports. On top of the reactor the gas-cooler is attached.

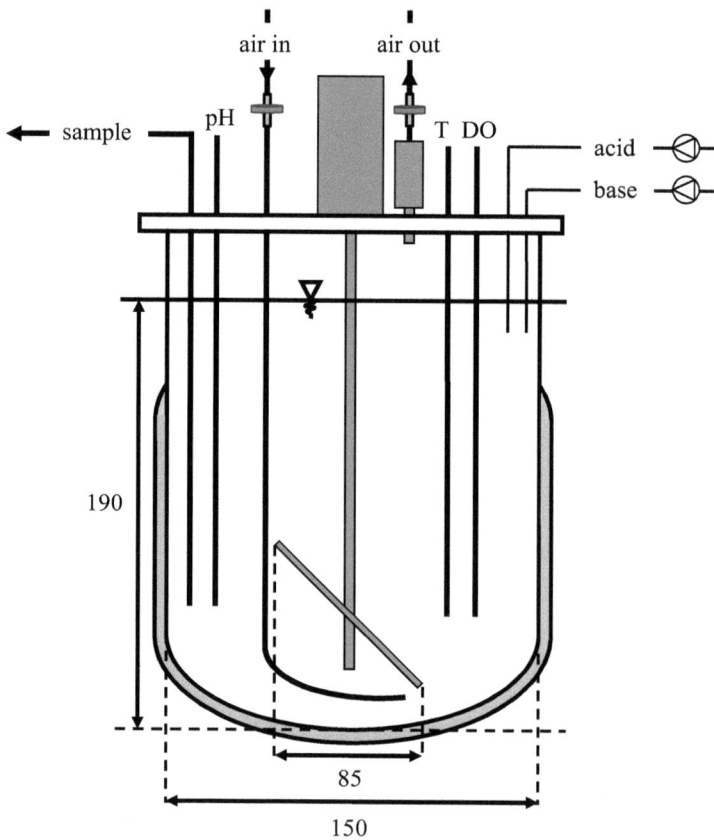

Figure 38 Schematic view of the Labfors® III stirred tank reactor - all dimensions in mm.

Chapter 4 - Development of a Novel Taylor-Couette Bioreactor

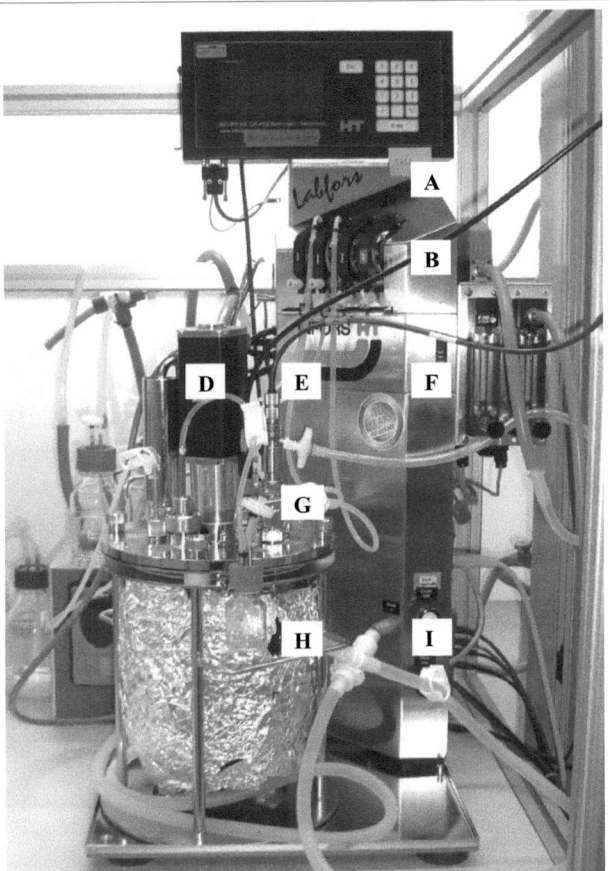

*Figure 39 Set-up of the STR. **A** Control unit with monitoring; **B** Pumps; **C** Gas cooler; **D** Motor; **E** Probes (pH, DO, T); **F** Gas inlet valve; **G** Feed, harvest lines; **H** Sampling device; **I** Temperature control.*

4.1.6 Numerical Details of the CFD Simulations

To evaluate the hydrodynamic stresses in the two bioreactors, CFD simulations of the flow field in both units are performed using FLUENT v6.2. Due to the vessel symmetry all simulations for the LTC are carried out using only one half of the unit height. The computation grid contained 20 nodes in the radial direction, 170 nodes in axial direction, and 420 nodes in the circumferential direction. In addition, points in the gap are distributed asymmetrically in order to properly resolve the local large gradients near the inner and outer walls. Thus, the total number of the computational cells (for 1/2 of the reactor height) was around 1'428'000. To take the turbulent flow in the LTC into account, a Reynolds Stress Model (FLUENT 6.2, 2005) combined with a standard wall function is employed. The rotation of the inner cylinder is implemented by a rotating reference frame where the liquid rotates with the same velocity as the inner cylinder, and the outer cylinder is treated as static (Sengupta et al., 2001).

In the case of the STR it was shown that the standard k-ε model (FLUENT 6.2, 2005) combined with a standard wall function is a suitable choice to characterize the flow field (Campolo et al., 2003; Kumaresan and Joshi, 2006; Soos et al., 2007). To simulate the rotation of the impeller the sliding mesh approach is adopted. The computational mesh contained approximately 500'000 elements.

4.2 Results and Discussion

4.2.1 Hydrodynamics and Flow Field Characterization

Since the STR as well as the LTC is not equipped with a torque meter and the Power numbers are not available, the evaluation of the energy dissipation rate is done using CFD simulations. As discussed earlier the cell damage during cultivation can be caused by hydrodynamic stress originating from gradients of the mean velocity, τ_L (Eq. 13), or from gradients of the fluctuating velocity field caused by turbulence, τ_{vs} (Eq. 19), or from the dynamic pressure acting on the opposite sides of the cell, τ_{IS} (Eq. 22). In order to distinguish whether cell damage occurs in the viscous subrange (Eq. 19) or in the inertial subrange (Eq. 22), the cell size is compared to the size of the smallest turbulent eddies characterized by the Kolmogorov microscale, κ (Eq. 18). The hydrodynamic stresses above are computed through CFD and the obtained values are compared with the experimentally determined threshold value ranging from 1 to 10 Pa above the cells undergo substantial damage (3.2.4). It is found that the LTC has to be operated below 75 rpm, corresponding to a volume average energy dissipation rate of 0.021 m^2 s^{-3}, in order to keep the hydrodynamic stress below the threshold value mentioned. Moreover, by comparing the size of the cells to the size of the smallest turbulent eddies characterized by the Kolmogorov microscale κ, it was found that for this operating conditions the cell size remains equal or smaller than κ. Consequently, any cell damage due to turbulence will occur in the viscose subrange and the corresponding hydrodynamic stress to which cells are exposed is evaluated using Eq. 19. A comparison of the distributions of τ_L and τ_{vs} evaluated from CFD simulation of the LTC operated at 75 rpm is shown in *Figure 40*.

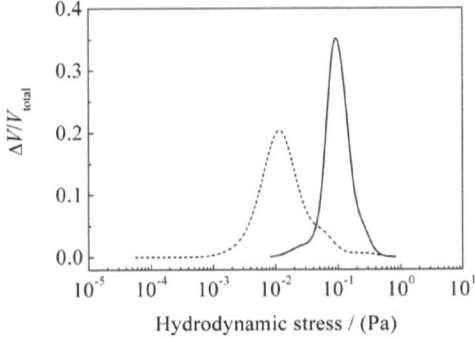

Figure 40 Comparison of the distribution of the hydrodynamic stresses originating from mean flow (---) and from turbulence in the viscous subrange (—) in the LTC device calculated for rotation speed of 75 rpm.

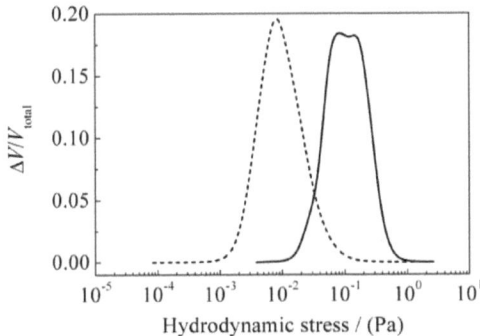

Figure 41 Comparison of the distribution of the hydrodynamic stresses originating from mean flow (---) and from turbulence in the viscous subrange (—) in the STR device calculated for rotation speed 130 rpm.

Chapter 4 - Development of a Novel Taylor-Couette Bioreactor

It can be concluded that turbulence in the viscous subrange is mostly responsible for cell damage. However, under those conditions the maximum value of τ_{vs} is around 0.8 Pa and therefore smaller than the threshold value leading to substantial cell damage (3.2.4).

In order to compare the performance of both bioreactors, the STR process is designed to operate at a similar volume average energy dissipation rate $\langle \varepsilon \rangle = 0.036$ m^2 s^{-3}, which corresponds to a rotation speed of the impeller equal to 130 rpm.

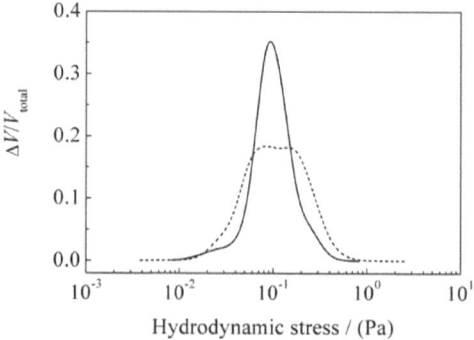

Figure 42 Distribution of the hydrodynamic stress gradients of the fluctuating velocity field caused by turbulence for the LTC (—) and for STR (---).

From the data shown in *Figure 41*, it is seen that, similar to the LTC, also for the STR the cell damage is solely controlled by the hydrodynamic stresses due to gradients of the fluctuating velocity field in the viscous subrange caused by turbulence. A comparison of τ_{vs} for both stirring devices is presented in *Figure 42*. It is seen, that the τ_{vs} distribution in the STR is substantially broader than in the LTC and covers three orders of magnitude. This goes from values around 2.7 Pa characteristic of the vicinity of the impeller blades to values around 0.004 Pa found at the periphery of the vessel, distant from the impeller. Suspended particles will therefore encounter two distinct hydrodynamic regions in an alternating fashion: a) high hydrodynamic stress regions in the vicinity of the stirring device b) low hydrodynamic stress regions away from the impeller. Comparing the values of the τ_{vs} maxima obtained in both devices, it is seen that the STR exhibits approximately three times larger hydrodynamic stresses compared to the LTC, although the average energy dissipation rate is similar.

It can be concluded that suspended cells in the LTC will experience a more homogenous microenvironment, without regions of high hydrodynamic stress, while traveling through the liquid. Additionally, zones of very low hydrodynamic stress, where mass transfer is inevitably limited, are avoided.

4.2.2 Mixing Time

To compare the performance of the STR and the LTC, the mixing time for both reactors are determined as in section 2.5 described. In *Figure 43* the dimensionless mixing time, Nt_M, representing the number of revolutions needed to achieve a certain degree of homogeneity, obtained by multiplying the mixing time with the rotation speed, is presented. For the STR it can be seen that the electrode placed in the bulk close to the impeller reacts faster than the electrode close to the liquid surface. Especially, under low rotation speeds the mixing process in the periphery of the reactor can take 60 % longer than in the bulk fluid. At higher agitation speed the difference becomes negligible. Since, the shear sensitivity of mammalian cells restricts the agitation speed to low values this circumstance may influence the homogeneity of the culture environment. For the LTC it can be seen, that for both electrodes the mixing time for all conditioned is the same. It can be seen that the LTC reactor needs under its particular set-up circa 1000 revolutions to achieve the desired homogeneity of 95 %. The STR is performing similar in approximately 50 revolutions. Nevertheless, it can be seen that Nt_M in the LTC is a constant over the range investigated, meaning the reactor is always working under turbulent conditions (Nagata, 1975). In contrary, Nt_M in the STR decreases at lower rotation speed, meaning the flow becomes transient or even laminar, where mass transfer limitations become unavoidable. Again, since the cultivation of mammalian cells has to be performed at low rotation speed, this circumstance becomes crucial.

Although, the mixing time within the STR is approximately one order of magnitude faster than in the LTC, the time required for reaching homogeneity in the LTC can be regarded as adequate performance for cell culturing. This is due to the fact that the time scales for mammalian cell growth and substrate consumption are in the range of hours and therefore rather kinetically than transport limited.

By comparing these results with other bubble-free reactor systems like the Wave® bioreactor, where typical mixing times of approximately 180 seconds are reported (Hearle et al., 2002), the LTC shows competitiveness. A summary of measured mixing times in various bioreactors is given elsewhere (Lara et al., 2006), indicating values ranging from a few seconds to a couple of minutes.

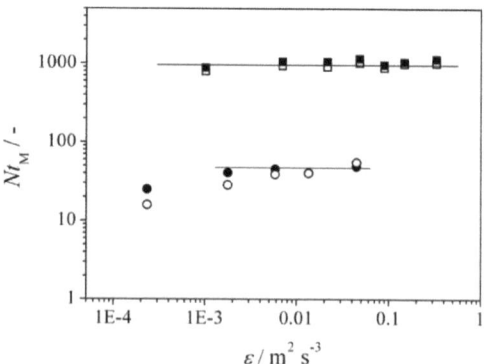

Figure 43 The dimensionless mixing time Nt_M as a function of the energy dissipation rate. (●) STR at the liquid surface; (○) STR in the bulk; (□,■) LTC at both reactor ends.

4.2.3 Mass transfer characteristics

The mass transfer coefficient for all units is determined by the dynamic method as described in 2.4. An example of the plot $\ln(1-y(t))$ versus time, to obtain $k_L a$ for the LTC at three rotation speeds (25, 50, and 100 rpm) using a transmembrane pressure (TMP) equal to 0.8 bar, is presented in *Figure 44*. Similar curves are determined for all conditions and both reactors. According to Eq. 7, the $k_L a$ values are obtained from the slopes of the fitted line to the experimental data as illustrated. A summary of the obtained $k_L a$ values for both reactors is presented in *Figure 45* and *Table 4*. It can be seen, that by employing higher agitation rates, turbulence increases and the liquid boundary film thickness is decreasing, leading to larger oxygen mass transfer. A similar effect on the mass transfer coefficient can be obtained by increasing the TMP and the aeration rate.

Chapter 4 - Development of a Novel Taylor-Couette Bioreactor

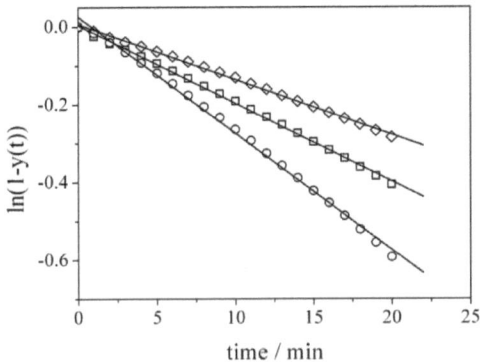

Figure 44 $k_L a$ determination (LTC data at 0.8 bar TMP). (◇) 25 (□) 50 (○) 100 rpm.

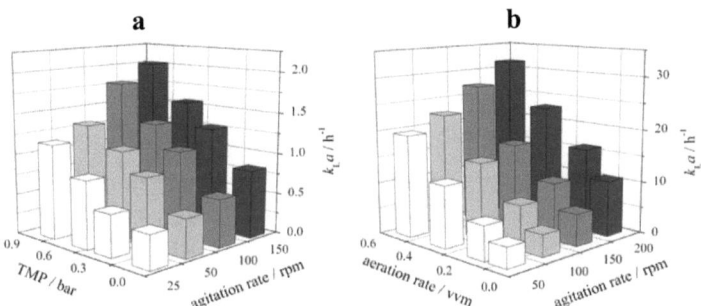

*Figure 45 Mass transfer coefficients for (**a**) the LTC as a function of the agitation rate and the TMP; (**b**) the STR as a function of the agitation and aeration rate.*

Comparing the values measured in this study with data published in the literature (see *Table* 3), it can be seen that the mass transfer properties of the LTC reactor are comparable with already established systems using membrane aeration, i.e. 1 - 5 h^{-1}. On the other hand, $k_L a$ values in sparged systems can reach values up to 30 h^{-1} and are up to one order of magnitude higher than bubble-free systems.

Considering that a mass transfer coefficient of around 1.0 h^{-1} is sufficient to ensure oxygen supply to cultured animal cells at densities up to 10^7 cells mL^{-1} (Fenge et al., 1993), from the data in *Table 4* it is seen that such a result could be obtained by operating the LTC reactor at 50 rpm and 0.5 bar inner cylinder pressure or at 100 rpm and 0.3 bar pressure.

	LTC					STR			
TMP rpm	0.0	0.3	0.5	0.8	vvm rpm	0.09	0.17	0.34	0.52
25	0.41	0.51	0.83	1.16	50	3.86	6.49	11.75	19.53
50	0.47	0.84	1.09	1.33	100	4.13	8.31	14.53	22.38
100	0.58	1.06	1.35	1.80	150	6.00	10.87	16.70	27.17
150	0.83	1.27	1.56	2.02	200	10.63	16.00	22.91	31.52

Table 4 Experimental $k_L a$ values (h^{-1}) for both reactors.

4.2.4 Membrane Lifetime and Reusability

To evaluate the mass transfer properties and reusability of the membrane material over various sterilization cycles, a scale-down model system is developed and tested. For this purpose a metal cassette with an installed membrane sheet (effective membrane area of 17.6 cm^2) is introduced into the STR. The vessel is filled with tap water, heated to 37° C and deoxygenized by flushing with nitrogen for 15 min. During this period also the head space of the reactor is thoroughly depleted of residual oxygen and therefore oxygen introduction via surface aeration is entirely avoided. The agitation rate is set to 100 rpm and an air pressure of 1.0 bar applied to the cassette. After recording the dissolved oxygen evolution until approximately 4 % air saturation the cassette is dismounted and autoclaved for 20 min at 121° C. This procedure is repeated six times and no effect of the sterilization on the membrane gas permeability found (*Figure 46*). It shall be noted that the slow oxygenation of the water derives from the small area/volume ratio of 0.66 and not a poor permeability of the membrane. Visual inspection of the membrane sheets did also not show any effect of the sterilization procedure.

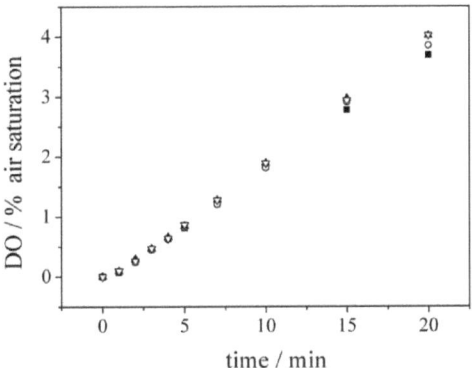

Figure 46 Membrane permeability after various sterilization cycles. (■) new membrane (○) after 1 (▲) after 3 (▽) after 6 sterilization cycles.

4.2.5 Cell Cultivation

Two cultivations systems, one using the LTC and the other the STR, are prepared as described in 4.1.4 and 4.1.5. According to the CFD analysis and the obtained threshold value for cell damage (3.2.4), the rotation speed is set to 75 rpm for the LTC resulting in a volume average energy dissipation rate of 0.021 $m^2 \, s^{-3}$ and a maximum hydrodynamic stress around 0.8 Pa. The same cultivation is performed in the STR while keeping a similar volume average energy dissipation rate, resulting in an agitation rate of 130 rpm for which the maximum value of τ_{vs} is around 2.7 Pa.

The inoculums for both reactors are prepared in roller bottles by feeding cells at a concentration of 3.5 x 10^5 cells mL^{-1} in fresh HTS medium 3-4 days before the start of the reactor cultivation. 24 hours before reactor inoculation, 20 % of the growth media in the roller bottles is substituted by fresh medium. At the same time 0.1 % (w/v) Pluronic F-68 is dissolved in 50 mL fresh HTS media, sterile filtrated and incubated at 37° C. The following day the F-68 solution is mixed with HTS media and pumped into the respective reactor. The LTC is mounted inclined (circa 5°) to ensure that residual gas is pushed out through the gas outlet. After reaching the predetermined set points for pH, DO and temperature, the inoculum (between 6-10 % of total reactor volume) is fed to the reactor. Liquid samples are taken on a regular basis. The solution is centrifuged at 18 g for 4 minutes and the supernatant aliquoted and frozen for further analysis at -20° C.

As mentioned earlier, any bioprocess has to be tightly controlled to maintain an optimum culture environment. A typical profile of temperature, pH, and DO during the cultivation in the LTC is shown in *Figure 47*. Obviously, the delivery of a constant temperature and pH can be easily achieved by the proposed design. Nevertheless, the DO value decreases over the culture time due to cell growth. By increasing the gas inlet pressure the system is able to maintain a DO > 20 % air saturation until 96 hours. After that point in time, the cell density and oxygen consumption is too high and the system can not sustain a sufficient oxygen delivery. This limitation can be resolved by feeding oxygen enriched air into the inner cylinder instead of pure air. By this simple adjustment, the aeration system should be able to deliver up to 5 fold more oxygen to the culture medium. Another point of concern is the selectivity of the used membrane. Up to now, it is not known whether all air components pass the silicone membrane at the same rate. If, for instance, nitrogen is slightly retained in the inner cylinder, it will accumulate and lead to a decreasing oxygen mass transfer. Therefore, the system shall be partially reconstructed to allow a constant gas exchange within the inner cylinder. It is worth noting that neither DO nor pH or temperature oscillations are observed during the STR run.

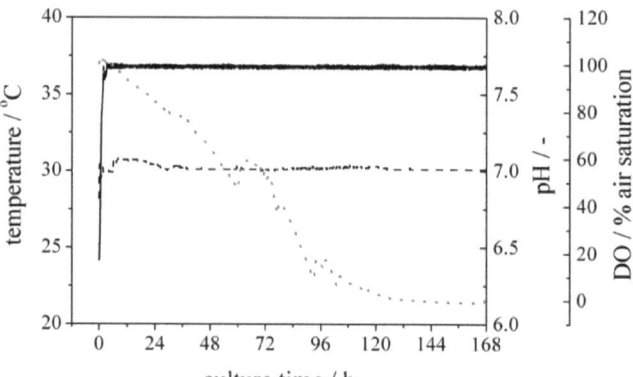

Figure 47 Stability and controllability of the LTC system during cultivation. (—) *Temperature* (•••) *pH* (•••) *DO.*

As can be seen in *Figure 48a,b* the LTC culture exhibits the typical growth profile with a short lag phase followed by exponential growth and finally reaching a plateau. The maximum cell density is equal to 1.8 x 10^6 cells mL^{-1} with a cell viability of 95 %. In contrast, the cell concentration and viability in the STR drops immediately after inoculation. These results support the CFD analysis combined with the experimentally obtained threshold value (3.2.4). The maximum value of the hydrodynamic stress in the LTC (0.8 Pa) is in fact below the threshold value (1 to 10 Pa) while in the STR the peak value (2.7 Pa) exceeds this limit.

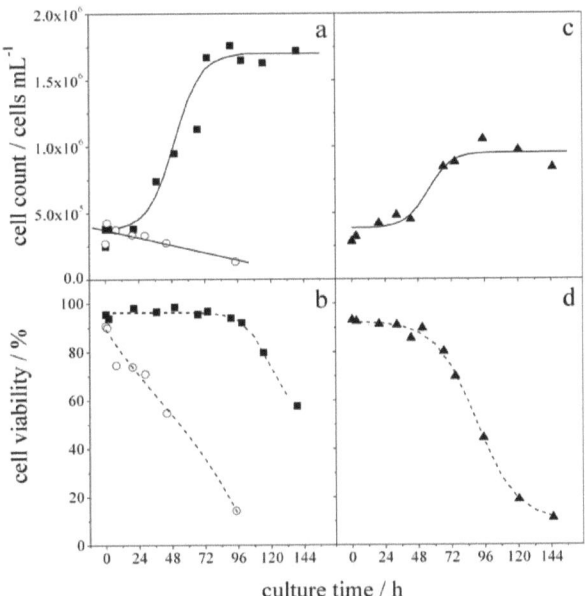

Figure 48 CHO cultivation in the (■) LTC and the (○) STR. (a, b) at a comparable energy dissipation rate (0.021 and 0.036 m^2 s^{-3}) (c, d) (▲) STR at an energy dissipation rate equal to 0.002 m^2 s^{-3}.

In order to reduce the hydrodynamic stresses in the STR the agitation rate was decreased to 50 rpm for which the volume averaged energy dissipation rate was equal to 0.002 m^2 s^{-3} and the corresponding maximum value of the hydrodynamic stress 0.8 Pa. We note that under this low stirring speed according to Langheinrich et al. (2002) the energy dissipated by aeration is approximately 1000 times smaller compared to the energy dissipation rate near the impeller. In addition, it was shown (see section 4.2.2) that under this rotation speed, mass transfer limitations are inevitable. The obtained growth profile is presented in *Figure 48c,d*. As it can be seen the growth profile for these conditions is similar to that obtained in the LTC, although the maximum cell density in the STR is smaller, i.e. about 1.1×10^6 cells mL^{-1}. A comparison of the cell growth rate in time for the two reactors is shown in *Figure 49*.

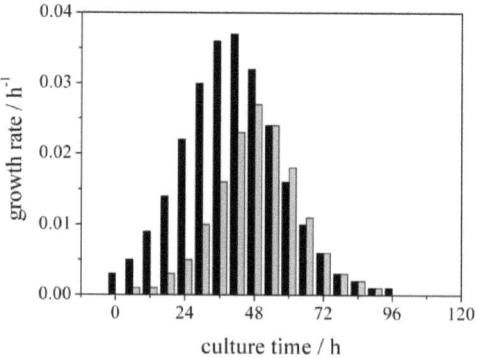

Figure 49 Growth rate of CHO cells in the different culture systems. Black bars: LTC culture, grey bars: STR culture.

It can be seen that the LTC culture is growing faster and exhibits a shorter lag phase. These results are in good agreement with studies made for the bioreactor cultivation of stem cells, where it was found, that the doubling time as well as the maximum cell density are significantly affected by the maximum shear stress occurring in the culture environment (Youn et al., 2006). The calculated doubling times are equal to 18.8 hours for the LTC culture and 25.9 hours for the STR. Based on these results it can be concluded that the higher energy input, leading to enhanced mass transfer, combined with the more homogenous microenvironment in the LTC reactor enables higher cell densities and faster growth.

As described in section 2.2.2 the metabolism of the cultivated cells can be retraced by HPLC analysis of the supernatant. In *Figure 50* a typical RI chromatogram displaying the nutrient glucose (eluting at 10.9 min) and the waste product lactate (eluting at 15.5 min) over the culture time, is shown. The graph shows exemplarily, the process of substrate consumption and by-product accumulation.

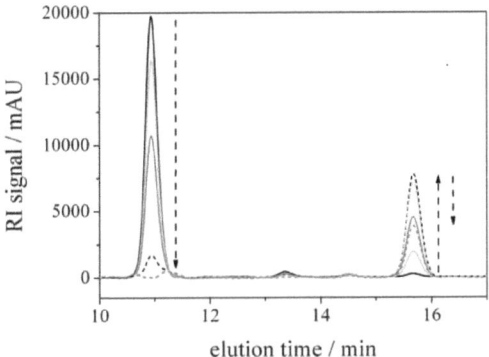

Figure 50 LTC run at 75 rpm with F-68. RI signal of liquid samples taken after (—) 0.2 (—) 20 (—) 50 (•••) 74 (•••) 99 (•••) 139 hours.

The glucose concentration decreases quickly and reaches zero after around 75 hours. In contrast, lactate accumulates until the glucose is completely consumed. At that point of time, the cells change their carbon source and start to metabolize also lactate. Hence, its liquid concentration drops. In *Figure 51*, the metabolic profile, in terms of the glucose consumption and lactate production, is illustrated. As expected these curves reflect the cell growth (*Figure 48*). In particular, at the point where the glucose concentration within the LTC medium reaches zero (after circa 75 hours), cell growth is slowing down and eventually stops. Due to a metabolic shift from glucose to lactate consumption (Messi, 2008), the lactate concentration starts to decrease after the glucose is depleted. A similar observation about the metabolic profile can be made for the STR run, with the exception that in this case the glucose concentration does not reach zero. This can be explained by looking at the culture viability profile (*Figure 48*), where the cell viability in the STR dropped to 50 % after 3.5 days resulting in less substrate consumption.

In other words, the cells in the LTC exhibit a higher metabolic activity and consume all the available glucose. In contrary, the STR culture is debilitated by the hydrodynamic environment and dies before able to consume all substrate.

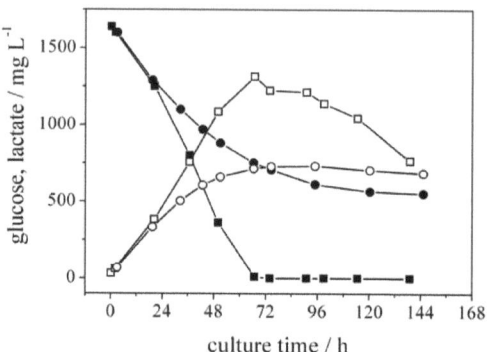

Figure 51 Comparison of the metabolic profile in the two culture systems (squares) LTC and (circles) STR. (○,□) Lactate and (●,■) Glucose concentrations.

To evaluate the effect of the hydrodynamic environment on the cells productivity, the SEAP concentration in the culture medium is determined as shown in *Figure 52a*. SEAP production in the LTC reactor reached its maximum after 4.5 days and leveled of at a value of 47 U L^{-1}. This corresponds to an increase, compared to the STR operated at $\langle \varepsilon \rangle$ = 0.002 m^2 s^{-3}, of almost 60 %, which can be attributed to the higher cell density and faster growth in the LTC. The resulting volumetric productivity is equal to 0.36 and 0.26 U L^{-1} h^{-1} for the LTC and the STR, respectively, indicating an increase of almost 40 % in the case of the LTC reactor. It is worth noting that this circumstance could be further exploited by implementing a biphasic growth concept, where the culture is first driven to high cell densities and subsequently growth-arrested to enhance productivity (Fussenegger et al., 1997; Mazur et al., 1998).

Another point of interest during cell cultivation for recombinant protein production emerges by looking at the purity of the resulting supernatant represented for instance by the amount of total protein released into the culture fluid. *Figure 52b* shows the profile of total released protein during the cultivation. Since the shear stress originating from agitation is below the threshold value for cell damage (3.2.4), the sharp increase in total protein concentration observed shortly after inoculation shows the detrimental effect of the bubble aeration employed in the STR. In fact, remarkable foam formation is observed as shown in *Figure 53*.

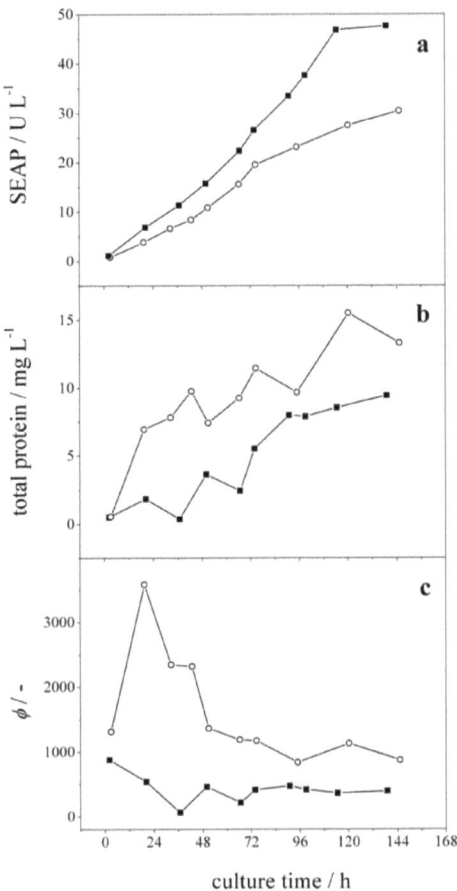

Figure 52 Comparison of the two culture systems (■) LTC and (○) STR. (a) SEAP production; (b) total protein accumulation in the supernatant; (c) contamination index ϕ.

Rising bubbles will transport suspended cells to the liquid surface and lead inevitable to an accumulation of cells within the foam layer. Moreover, during the bursting of the bubbles, severe cell damage can be expected, leading to a decreased cell viability and a loss in productivity (Cherry and Hulle, 1992; Handa-Corrigan et al., 1989; Michaels et al., 1996). By the implementation of the membrane aeration system in the zero-head space LTC reactor, these kinds of problems are eliminated. The LTC culture shows no significant damage and the release of protein into the medium is linearly increasing over the culture time. The contamination index ϕ (Eq. (5)) relating the amount of total protein and SEAP within the supernatant is presented in *Figure 52c*. As expected, the initial damage of the weakest cells after inoculation leads to large values of the ϕ-ratio in the beginning of the STR process. During cultivation, the contamination index decreases due to the SEAP production. Nevertheless, at the end of the cultivation, the amount of contaminating host cell proteins (HCP) in the STR supernatant exceeds the LTC value by a factor of two, making the downstream processing of the STR product more complicated and expensive.

Figure 53 Foaming due to bubble aeration during cell cultivation in the STR (at 50 rpm, 0.1 vvm, with protein-free HTS media supplemented with 0.1 w/v % F-68).

It is also worth noting that, damaged cells are known to release other undesired species as for instance sialidases, which can affect the oligosaccharide structure of the produced glycoproteins (Goldman et al., 1997; Gramer and Goochee, 1993) and hence the product yield. Further optimization of the LTC process could also lead to a reduced harvest time and therefore minimize the concentration of contaminants and proteases within the supernatant.

Finally, the dependency of cell growth on Pluronic F-68 in both systems is evaluated. Both reactors were operated at comparable maximum hydrodynamic stresses under the conditions described above without the supplementation of F-68. Cell cultivation performed in the STR at 50 rpm and 0.1 vvm using F-68 free medium led to an immediate decline in the viable cell density resulting in a dead culture after 36 hours (data not shown). In contrary, the LTC reactor permits the cultivation of mammalian cells without the addition of any cell protective agents (data not shown) and reached cell densities at least equal to that in the STR with F-68. By avoiding the medium supplement Pluronic F-68, mass transfer is enhanced (Kawase and Mooyoung, 1990) and downstream processing simplified.

4.3 Conclusions

The major objective of this study was to develop a novel bioreactor, combining the benefits of existing reactor systems, to establish a reactor unit suitable for the cultivation of shear sensitive cells and tissues. It was shown that the proposed device is featuring good mass transfer characteristics at low shear values both in maximal magnitude as well as distribution. Further the lobed Taylor-Couette bioreactor emerged superior to a common STR regarding cell cultivation. It was proven that the lag phase after inoculation is shortened, the maximum cell density as well as the growth rate is increased and that the volumetric productivity is enhanced. Moreover, the LTC process introduces ten times more energy to the culture fluid, compared to the STR, without sacrificing cell viability and productivity. By implementing a zero-head-space, the working volume is increased and foaming, leading to the accumulation and loss of medium components or cells, is avoided. Moreover, cell cultivation without the supplementation with cell protective and antifoam reagents is feasible. Especially nowadays, where product integrity and consistency are of uttermost importance and strongly controlled by the regulatory authorities, the LTC provides a suitable tool for biopharmaceutical research, development and production. Possible applications could vary from culturing cell suspensions and cell aggregates, large plant cells, adherent cells on microcarriers, shear-sensitive biocatalysts, protoplasts and stem cells to tissue scaffolds.

Chapter 5
Conclusions

A detailed study of the response of mammalian cells to environmental stress, namely shear stress, has been conducted in this work. Based to these results, a novel bioreactor was constructed in order to overcome the limitations of established culture systems.

In the first part of this work, it was shown that mammalian cells respond differently when exposed to steady simple shear or extensional flow patterns. Exceeding a critical threshold value, the cultures will either enter the apoptotic pathway or dye by necrosis. In simple shear flow, low values of hydrodynamic stress cause severe cell necrosis. In contrast, when cells are exposed to extensional flow with hydrodynamic stresses of the same amplitude, operated under laminar conditions, a very short exposure to the hydrodynamic stress leads to the induction of the apoptotic death pathway. During short term exposure in extensional flow, necrosis was observed when the cell size was larger than the Kolmogorov length scale resulting in significantly larger values of the critical hydrodynamic stress with respect to the simple shear case. However, under prolonged exposure to extensional flow cells suffer severe necrosis even under shear stresses comparable to the threshold obtained using simple shear flow.

In order to avoid the hydrodynamic stresses larger than these threshold values a novel bioreactor for the cultivation of shear sensitive cells and tissues was developed. The main benefits of the proposed device compared to the well established stirred tank reactor were (a) a gentle hydrodynamic environment with low shear values both in maximal magnitude as well as distribution (b) a shortened lag phase after inoculation (c) a higher maximum cell density and growth rate (d) an increased volumetric productivity (e) an enhanced power input without sacrificing cell viability or productivity and (f) the avoidance of cell protective and antifoam reagents during cultivation.

Chapter 5 - Conclusions

The results obtained from the comparison of the culture systems proved the suitability of the in the first part of this thesis proposed model systems, to predict the critical level of shear stress for cell damage in real cultivation systems. By implementing the proposed experimental set-up for evaluating the threshold value, expensive and time consuming small-scale experiments in bioreactors to establish the operating conditions, namely the agitation rate, can be avoided.

Especially nowadays, where product integrity and consistency are of uttermost importance and strongly controlled by the regulatory authorities, the lobed Taylor-Couette bioreactor provides a suitable tool for biopharmaceutical research, development and production. Possible applications could vary from culturing cell suspensions and cell aggregates, large plant cells, adherent cells on microcarriers, shear-sensitive biocatalysts, protoplasts and stem cells to tissues grown on scaffolds.

Chapter 6
Outlook

The lobed Taylor-Couette bioreactor was developed in-house and proved to be a promising alternative to already established cell cultivation systems. Nevertheless, the LTC unit is still in its initial phase of development and requires further optimization. The following issues should be addressed in advanced studies:

Up to now the LTC unit is operated at an agitation speed of 75 rpm, equal to a volume average energy dissipation rate of 0.021 m^2 s^{-3}. This value was chosen, based on the results obtained during the cell damage experiments. Nevertheless, the critical threshold value for cell damage is still not accurately determined giving rise to the possibility of higher agitation speeds and therefore enhanced mass transfer. Moreover, the product quality and purity in both reactor types should be further investigated. Especially, the benefit of a higher viability in the LTC, connected to a decreased release of HCP and DNA, is of uttermost importance for the subsequent downstream-processing of produced proteins.

Another point of interest arises from the results obtained during cell cultivation. It was observed that glucose depletion occurs already after 72 hours of cultivation, leading to substrate limitation and a reduced cell viability. In this context, medium optimization and the development of a proper feeding strategy should be considered to enhance the culture longevity. Additionally, by the implementation of a continuous or perfusion operation mode, the productivity could be further increased. A perfusion system could be for instance accomplished by the installation of a by-pass channel connected to a filtration system conducting the cell retention (Buntemeyer et al., 1994; Castilho et al., 2002).

Moreover the process environment within the LTC unit, especially the dissolved oxygen concentration (DO), is not yet optimized. Due to the current oxygen delivery method, the system suffers from a non-constant DO over the cultivation time. Since, the

Chapter 6 - Outlook

dissolved oxygen tension in the culture medium has been shown to affect productivity, cell metabolism and protein glycosylation (Boise et al., 1993; Kunkel et al., 2000; Kunkel et al., 1998; Kurano et al., 1990; Link et al., 2004; Ozturk and Palsson, 1990), the optimization of the aeration system has to be addressed. Here, a dynamic adaptation of the gas pressure to the changing culture needs seems to be a promising solution. Additionally, the inner cylinder should be equipped with a gas outlet to avoid a potential selective gas accumulation within the pressurized space.

Even though the LTC reactor shows superior process performance in comparison to the STR, it will be difficult to substitute this well established system in the field of mammalian suspension cell cultivation. Nevertheless, there are other market niches like the 3D cultivation of mammalian tissues or stem cells, which lately draw more and more attention. Large-scale production of these kinds of cells is still difficult due to a lack of a suitable culture system with a robust micro-environmental control. Static cultures do not provide steady-state operating conditions and are limited in the cultivation area. Therefore, stirred suspension culture techniques were explored and evaluated regarding their suitability. Most of the proposed systems are modifications of the common stirred tank reactor, suffering from diverse drawbacks like a broad distribution of the energy dissipation rate. Here, the LTC reactor could fill the gap providing a more homogenous and controlled microenvironment supporting effectively cell proliferation by the mild and fine-tuned shear environment.

Chapter 7
Bibliography

Abu-Reesh I, Kargi F. 1989. Biological Responses of Hybridoma Cells to Defined Hydrodynamic Shear-Stress. Journal of Biotechnology 9(3):167-177.

Ahmed F, Alexandridis P, Neelamegham S. 2001. Synthesis and application of fluorescein-labeled Pluronic block copolymers to the study of polymer-surface interactions. Langmuir 17(2):537-546.

Al-Rubeai M, Emery AN, Chalder S. 1992. The Effect of Pluronic F-68 on Hybridoma Cells in Continuous Culture. Applied Microbiology and Biotechnology 37(1):44-45.

Al-Rubeai M, Singh RP, Goldman MH, Emery AN. 1995. Death Mechanisms of Animal-Cells in Conditions of Intensive Agitation. Biotechnology and Bioengineering 45(6):463-472.

Alexopoulos AH, Maggioris D, Kiparissides C. 2002. CFD analysis of turbulence non-homogeneity in mixing vessels - A two-compartment model. Chemical Engineering Science 57(10):1735-1752.

Aloi LE, Cherry RS. 1994. Intracellular Calcium Response of Sf-9 Insect Cells Exposed to Intense Fluid Forces. Journal of Biotechnology 33(1):21-31.

Aloi LE, Cherry RS. 1996. Cellular response to agitation characterized by energy dissipation at the impeller tip. Chemical Engineering Science 51(9):1523-1529.

Alvarez MM, Guzman A, Elias M. 2005. Experimental visualization of mixing pathologies in laminar stirred tank bioreactors. Chemical Engineering Science 60(8-9):2449-2457.

Chapter 7 - Bibliography

Andereck CD, Liu SS, Swinney HL. 1986. Flow Regimes in a Circular Couette System with Independently Rotating Cylinders. Journal of Fluid Mechanics 164:155-183.

Andersen DC, Goochee CF. 1995. The Effect of Ammonia on the O-Linked Glycosylation of Granulocyte-Colony-Stimulating Factor Produced by Chinese-Hamster Ovary Cells. Biotechnology and Bioengineering 47(1):96-105.

Anderson CD, Schwarz RP; Synthecon Inc, assignee. 1998. Bioreactor for culturing cells and tissues in suspension in liquid nutrient|made partially of gas permeable materials with ports for access to culture, and rotated to suspend cells and tissues without turbulence patent US5763279-A.

Andree HAM, Reutelingsperger CPM, Hauptmann R, Hemker HC, Hermens WT, Willems GM. 1990. Binding of Vascular Anticoagulant-Alpha (Vac-Alpha) to Planar Phospholipid-Bilayers. Journal of Biological Chemistry 265(9):4923-4928.

Apenberg S, Freyberg MA, Friedl P. 2003. Shear stress induces apoptosis in vascular smooth muscle cells via an autocrine Fas/FasL pathway. Biochemical and Biophysical Research Communications 310(2):355-359.

Arratia PE, Lacombe JP, Shinbrot T, Muzzio FJ. 2004. Segregated regions in continuous laminar stirred tank reactors. Chemical Engineering Science 59(7):1481-1490.

Augenstein DC, Sinskey AJ, Wang DIC. 1971. Effect of Shear on Death of Two Strains of Mammalian Tissue Cells. Biotechnology and Bioengineering 13(3):409-418.

Bandyopadhyay B, Humphrey AE, Taguchi H. 1967. Dynamic Measurement of Volumetric Oxygen Transfer Coefficient in Fermentation Systems. Biotechnology and Bioengineering 9(4):533-544.

Bavarian F, Fan LS, Chalmers JJ. 1991. Microscopic Visualization of Insect Cell Bubble Interactions .1. Rising Bubbles, Air Medium Interface, and the Foam Layer. Biotechnology Progress 7(2):140-150.

Beeton S, Millward HR, Bellhouse BJ, Nicholson AM, Jenkins N, Knowles CJ. 1991. Gas Transfer Characteristics of a Novel Membrane Bioreactor. Biotechnology and Bioengineering 38(10):1233-1238.

Bennett MR, Gibson DF, Schwartz SM, Tait JF. 1995. Binding and Phagocytosis of Apoptotic Vascular Smooth-Muscle Cells Is Mediated in Part by Exposure of Phosphatidylserine. Circulation Research 77(6):1136-1142.

Bentley BJ, Leal L. 1986. An experimental investigation of drop deformation and breakup in steady, two-dimensional linear flows. Journal of Fluid Mechanics 167:241-283.

Bentley PK, Gates RMC, Lowe KC, Depomerai DI, Walker JAL. 1989. Invitro Cellular-Responses to a Non-Ionic Surfactant, Pluronic F-68. Biotechnology Letters 11(2):111-114.

Bird RB, Stewart WE, Lightfoot EN. 2002. Transport Phenomena. New York: John Wiley&Sons, Inc.

Blaser S. 1998. The hydrodynamical effect of vorticity and strain on the mechanical stability of flocs [PhD thesis No. 12851]. Zurich: ETH Zurich.

Blaser S. 2000a. Break-up of flocs in contraction and swirling flows. Colloids and Surfaces A 166:215-223.

Blaser S. 2000b. Flocs in shear and strain flows. Journal of Colloid and Interface Science 225:273-284.

Bohme C, Schroder MB, JungHeiliger H, Lehmann J. 1997. Plant cell suspension culture in a bench-scale fermenter with a newly designed membrane stirrer for bubble-free aeration. Applied Microbiology and Biotechnology 48(2):149-154.

Boise LH, Gonzalezgarcia M, Postema CE, Ding LY, Lindsten T, Turka LA, Mao XH, Nunez G, Thompson CB. 1993. Bcl-X, a Bcl-2-Related Gene That Functions as a Dominant Regulator of Apoptotic Cell-Death. Cell 74(4):597-608.

Chapter 7 - Bibliography

Born C, Zhang Z, Alrubeai M, Thomas CR. 1992. Estimation of Disruption of Animal-Cells by Laminar Shear-Stress. Biotechnology and Bioengineering 40(9):1004-1010.

Borys MC, Linzer DIH, Papoutsakis ET. 1993. Culture Ph Affects Expression Rates and Glycosylation of Recombinant Mouse Placental-Lactogen Proteins by Chinese-Hamster Ovary (Cho) Cells. Bio-Technology 11(6):720-724.

Bouyer D, Coufort C, Line A, Do-Quang Z. 2005. Experimental analysis of floc size distributions in a 1-L jar under different hydrodynamics and physicochemical conditions. Journal of Colloid and Interface Science 292(2):413-428.

Bradford MM. 1976. Rapid and Sensitive Method for Quantitation of Microgram Quantities of Protein Utilizing Principle of Protein-Dye Binding. Analytical Biochemistry 72(1-2):248-254.

Brennan DJ, Lehrer IH. 1976. Impeller Mixing in Vessels - Experimental Studies on Influence of Some Parameters and Formulation of a General Mixing Time Equation. Transactions of the Institution of Chemical Engineers 54(3):139-152.

Buntemeyer H, Bohme C, Lehmann J. 1994. Evaluation of Membranes for Use in Online Cell-Separation During Mammalian-Cell Perfusion Processes. Cytotechnology 15(1-3):243-251.

Butler M. 2005. Animal cell cultures: recent achievements and perspectives in the production of biopharmaceuticals. Applied Microbiology and Biotechnology 68(3):283-291.

Campolo M, Sbrizzai F, Soldati A. 2003. Time-dependent flow structures and Lagrangian mixing in Rushton-impeller baffled-tank reactor. Chemical Engineering Science 58(8):1615-1629.

Castilho LR, Anspach FB, Deckwer WD. 2002. An integrated process for mammalian cell perfusion cultivation and product purification using a dynamic filter. Biotechnology Progress 18(4):776-781.

Chapter 7 - Bibliography

Cervantes MIS, Lacombe J, Muzzio FJ, Alvarez MM. 2006. Novel bioreactor design for the culture of suspended mammalian cells. Part 1: Mixing characterization. Chemical Engineering Science 61(24):8075-8084.

Chattopadhyay D, Rathman JF, Chalmers JJ. 1995. The Protective Effect of Specific Medium Additives with Respect to Bubble Rupture. Biotechnology and Bioengineering 45(6):473-480.

Cherry RS, Hulle CT. 1992. Cell-Death in the Thin-Films of Bursting Bubbles. Biotechnology Progress 8(1):11-18.

Chisti Y. 2001. Hydrodynamic damage to animal cells. Critical Reviews in Biotechnology 21(2):67-110.

Cole JA. 1967. Taylor Vortices with Eccentric Rotating Cylinders. Nature 216(5121):1200-1202.

Couette M. 1890. Études sur le frottement des liquides. Annales de chimie et de physique 6(21):433-510.

Cowger NL, O'Connor KC, Hammond TG, Lacks DJ, Navar GL. 1999. Characterization of bimodal cell death of insect cells in a rotating-wall vessel and shaker flask. Biotechnology and Bioengineering 64(1):14-26.

Cowger NL, Oconnor KC, Bivins JE. 1997. Influence of simulated microgravity on the longevity of insect-cell culture. Enzyme and Microbial Technology 20(5):326-332.

Croughan MS, Wang DIC. 1989. Growth and Death in Overagitated Microcarrier Cell-Cultures. Biotechnology and Bioengineering 33(6):731-744.

Cruz HJ, Freitas CM, Alves PM, Moreira JL, Carrondo MJT. 2000. Effects of ammonia and lactate on growth, metabolism, and productivity of BHK cells. Enzyme and Microbial Technology 27(1-2):43-52.

Curran SJ, Black RA. 2004. Quantitative experimental study of shear stresses and mixing in progressive flow regimes within annular-flow bioreactors. Chemical Engineering Science 59(24):5859-5868.

Cutter LA. 1966. Flow and Turbulence in a Stirred Tank. Aiche Journal 12(1):35-45.

Darzynkiewicz Z, Bruno S, Delbino G, Gorczyca W, Hotz MA, Lassota P, Traganos F. 1992. Features of Apoptotic Cells Measured by Flow-Cytometry. Cytometry 13(8):795-808.

Darzynkiewicz Z, Juan G, Li X, Gorczyca W, Murakami T, Traganos F. 1997. Cytometry in cell necrobiology: Analysis of apoptosis and accidental cell death (necrosis). Cytometry 27(1):1-20.

Deshmukh M, Johnson EM. 2000. Staurosporine-induced neuronal death: multiple mechanisms and methodological implications. Cell Death and Differentiation 7(3):250-261.

Desmet G, Verelst H, Baron GV. 1996. Local and global dispersion effects in Couette-Taylor flow .1. Description and modeling of the dispersion effects. Chemical Engineering Science 51(8):1287-1298.

deZengotita VM, Schmelzer AE, Miller WM. 2002. Characterization of hybridoma cell responses to elevated pCO(2) and osmolality: Intracellular pH, cell size, apoptosis, and metabolism. Biotechnology and Bioengineering 77(4):369-380.

Di Prima RC, Swinney HL. 1979. Instabilities and transition in flow between concentric rotating cylinders In: Swinney HL, Gollub JP, editors. Hydrodynamic Instabilities and the Transition to Turbulence. Berlin / Heidelberg: Springer p139-180.

Dickson M, Gagnon JP. 2004. Key factors in the rising cost of new drug discovery and development. Nature Reviews Drug Discovery 3(5):417-429.

Dive C, Gregory CD, Phipps DJ, Evans DL, Milner AE, Wyllie AH. 1992. Analysis and Discrimination of Necrosis and Apoptosis (Programmed Cell-Death) by Multiparameter Flow-Cytometry. Biochimica Et Biophysica Acta 1133(3):275-285.

Chapter 7 - Bibliography

Ducommun P, Ruffieux PA, Furter MP, Marison I, von Stockar U. 2000. A new method for on-line measurement of the volumetric oxygen uptake rate in membrane aerated animal cell cultures. Journal of Biotechnology 78(2):139-147.

Dunlop EH, Namdev PK, Rosenberg MZ. 1994. Effect of Fluid Shear Forces on Plant-Cell Suspensions. Chemical Engineering Science 49(14):2263-2276.

Eagle H. 1955. Nutrition Needs of Mammalian Cells in Tissue Culture. Science 122(3168):501-504.

Einsele A, Ristroph DL, Humphrey AE. 1978. Mixing Times and Glucose-Uptake Measured with a Fluorometer. Biotechnology and Bioengineering 20(9):1487-1492.

Even MS, Sandusky CB, Barnard ND. 2006. Serum-free hybridoma culture: ethical, scientific and safety considerations. Trends in Biotechnology 24(3):105-108.

Ezra E, Blacher R, Udenfriend S. 1983. Purification and Partial Sequencing of Human Placental Alkaline Phosphatase Ec-3.1.3.1. Biochemical and Biophysical Research Communications 116(3):1076-1083.

Fadok VA, Voelker DR, Campbell PA, Cohen JJ, Bratton DL, Henson PM. 1992. Exposure of Phosphatidylserine on the Surface of Apoptotic Lymphocytes Triggers Specific Recognition and Removal by Macrophages. Journal of Immunology 148(7):2207-2216.

Fenge C, Klein C, Heuer C, Siegel U, Fraune E. 1993. Agitation, Aeration and Perfusion Modules for Cell-Culture Bioreactors. Cytotechnology 11(3):233-244.

Fleischaker RJ, Sinskey AJ. 1981. Oxygen-Demand and Supply in Cell-Culture. European Journal of Applied Microbiology and Biotechnology 12(4):193-197.

FLUENT 6.2. 2005. User's Guide.

Franek F, Fussenegger M. 2005. Survival factor-like activity of small peptides in hybridoma and CHO cells cultures. Biotechnology Progress 21(1):96-98.

Chapter 7 - Bibliography

Fussenegger M, Mazur X, Bailey JE. 1997. A novel cytostatic process enhances the productivity of Chinese hamster ovary cells. Biotechnology and Bioengineering 55(6):927-939.

Garciabriones MA, Chalmers JJ. 1994. Flow Parameters Associated with Hydrodynamic Cell Injury. Biotechnology and Bioengineering 44(9):1089-1098.

Geisse S, Kocher HP. 1999. Protein expression in mammalian and insect cell systems. Expression of Recombinant Genes in Eukaryotic Systems 306:19-42.

Goldman MH, James DC, Ison AP, Bull AT. 1997. Monitoring proteolysis of recombinant human interferon-gamma during batch culture of Chinese hamster ovary cells. Cytotechnology 23(1-3):103-111.

Goodwin TJ, Prewett TL, Wolf DA, Spaulding GF. 1993. Reduced Shear-Stress - a Major Component in the Ability of Mammalian-Tissues to Form 3-Dimensional Assemblies in Simulated Microgravity. Journal of Cellular Biochemistry 51(3):301-311.

Gorczyca W, Gong JP, Darzynkiewicz Z. 1993. Detection of DNA Strand Breaks in Individual Apoptotic Cells by the Insitu Terminal Deoxynucleotidyl Transferase and Nick Translation Assays. Cancer Research 53(8):1945-1951.

Gramer MJ, Goochee CF. 1993. Glycosidase Activities in Chinese-Hamster Ovary Cell Lysate and Cell-Culture Supernatant. Biotechnology Progress 9(4):366-373.

Gregoriades N, Clay J, Ma N, Koelling K, Chalmers JJ. 2000. Cell damage of microcarrier cultures as a function of local energy dissipation created by a rapid extensional flow. Biotechnology and Bioengineering 69(2):171-182.

Handa-Corrigan A, Emery AN, Spier RE. 1989. Effect of Gas-Liquid Interfaces on the Growth of Suspended Mammalian-Cells - Mechanisms of Cell-Damage by Bubbles. Enzyme and Microbial Technology 11(4):230-235.

Handacorrigan A, Emery AN, Spier RE. 1989. Effect of Gas-Liquid Interfaces on the Growth of Suspended Mammalian-Cells - Mechanisms of Cell-Damage by Bubbles. Enzyme and Microbial Technology 11(4):230-235.

Hearle DC, Osman J, Knevelman C, Khan M. 2002. Characterization and operation of a disposable bioreactor as a replacement for conventional steam-in-place inoculum bioreactors for mammalian cell culture processes. Abstracts of Papers of the American Chemical Society 224:U233-U233.

Heath C, Kiss R. 2007. Cell culture process development: Advances in process engineering. Biotechnology Progress 23(1):46-51.

Higashitani K, Inada N, Ochi T. 1991. Floc breakup along centerline of contractile flow to orifice. Colloids and Surfaces 56:13-23.

Hinze JO. 1955. Fundamentals of the Hydrodynamic Mechanism of Splitting in Dispersion Processes. Aiche Journal 1(3):289-295.

Hooker BS, Lee JM, An G. 1989. Response of Plant-Tissue Culture to a High Shear Environment. Enzyme and Microbial Technology 11(8):484-490.

Hvoslef-Eide A, Olsen O, Lyngved R, Munster C, Heyerdahl P. 2005. Bioreactor design for propagation of somatic embryos Plant Cell, Tissue and Organ Culture 81(3):265-276.

Ito M, McLimans WF. 1981. Ammonia Inhibition of Interferon Synthesis. Cell Biology International Reports 5(7):661-666.

Janes DA, Thomas NH, Callow JA. 1987. Demonstration of a Bubble-Free Annular-Vortex Membrane Bioreactor for Batch Culture of Red Beet Cells. Biotechnology Techniques 1(4):257-262.

Joosten CE, Shuler ML. 2003. Effect of culture conditions on the degree of sialylation of a recombinant glycoprotein expressed in insect cells. Biotechnology Progress 19(3):739-749.

Jordan M, Sucker H, Einsele A, Widmer F, Eppenberger HM. 1994. Interactions between Animal-Cells and Gas-Bubbles - the Influence of Serum and Pluronic F68 on the Physical-Properties of the Bubble Surface. Biotechnology and Bioengineering 43(6):446-454.

Kataoka K. 1986. Taylor vortices and instabilities in circular Couette flows. In: Cheremisinoff NP, editor. Encyclopedia of Fluid Mechanics. Houston: Gulf Pub. p 236–274.

Kawase Y, Mooyoung M. 1990. The Effect of Antifoam Agents on Mass-Transfer in Bioreactors. Bioprocess Engineering 5(4):169-173.

Keane JT, Ryan D, Gray PP. 2003. Effect of shear stress on expression of a recombinant protein by Chinese hamster ovary cells. Biotechnology and Bioengineering 81(2):211-220.

Kilburn DG, Morley M. 1972. Method for Controlling Dissolved-Oxygen Tension and Measuring Respiration Rate in Suspension Cultures of Animal Cells. Biotechnology and Bioengineering 14(3):499-504.

Kohler C, Orrenius S, Zhivotovsky B. 2002. Evaluation of caspase activity in apoptotic cells. Journal of Immunological Methods 265(1-2):97-110.

Koopman G, Reutelingsperger CPM, Kuijten GAM, Keehnen RMJ, Pals ST, van Oers MHJ. 1994. Annexin-V for Flow Cytometric Detection of Phosphatidylserine Expression on B-Cells Undergoing Apoptosis. Blood 84(10):1415-1420.

Kretzmer G. 2001. Influence of Stress on Adherent Cells In: Scheper T, editor. Advances in Biochemical Engineering/Biotechnology: Influence of Stress on Cell Growth and Product Formation. Berlin / Heidelberg: Springer

Kretzmer G, Schugerl K. 1991. Response of Mammalian-Cells to Shear-Stress. Applied Microbiology and Biotechnology 34(5):613-616.

Kumaresan T, Joshi JB. 2006. Effect of impeller design on the flow pattern and mixing in stirred tanks. Chemical Engineering Journal 115(3):173-193.

Kunas KT, Papoutsakis ET. 1989. Increasing Serum Concentrations Decrease Cell-Death and Allow Growth of Hybridoma Cells at Higher Agitation Rates. Biotechnology Letters 11(8):525-530.

Kunkel JP, Jan DCH, Butler M, Jamieson JC. 2000. Comparisons of the glycosylation of a monoclonal antibody produced under nominally identical cell culture conditions in two different bioreactors. Biotechnology Progress 16(3):462-470.

Kunkel JP, Jan DCH, Jamieson JC, Butler M. 1998. Dissolved oxygen concentration in serum-free continuous culture affects N-linked glycosylation of a monoclonal antibody. Journal of Biotechnology 62(1):55-71.

Kurano N, Leist C, Messi F, Kurano S, Fiechter A. 1990. Growth-Behavior of Chinese Hamster Ovary Cells in a Compact Loop Bioreactor .1. Effects of Physical and Chemical Environments. Journal of Biotechnology 15(1-2):101-111.

Kwon O, Devarakonda SB, Sankovic JM, Banerjee RK. 2008. Oxygen transport and consumption by suspended cells in microgravity: A multiphase analysis. Biotechnology and Bioengineering 99(1):99-107.

Laken HA, Leonard MW. 2001. Understanding and modulating apoptosis in industrial cell culture. Current Opinion in Biotechnology 12(2):175-179.

Langheinrich C, Nienow AW, Eddleston T, Stevenson NC, Emery AN, Clayton TM, Slater NKH. 2002. Oxygen transfer in stirred bioreactors under animal cell culture conditions. Food and Bioproducts Processing 80(C1):39-44.

Lara AR, Galindo E, Ramirez OT, Palomares LA. 2006. Living with heterogeneities in bioreactors. Molecular Biotechnology 34(3):355-381.

Lasheras JC, Eastwood C, Martínez-Bazán C, Montañés JL. 2002. A review of statistical models for the break-up of an immiscible fluid immersed into a fully developed turbulent flow. International Journal of Multiphase Flow 28:247-278.

Lawrence S. 2005. Biotech drug market steadily expands. Nature Biotechnology 23(12):1466-1466.

Lehmann J, Piehl GW, Schulz R. 1987. Bubble Free Cell Culture Aeration with Porous Moving Membranes. International Association of Biological Standardization (Ed.). Developments in Biological Standardization, Vol. 66. Advances in Animal Cell

Technology: Cell Engineering, Evaluation and Exploitation; 77th General Meeting of European Society for Animal Cell Technology, Vienna, Austria, September 30-October 4, 1985. Xii+581p. S. Karger Ag: Basel, Switzerland; New York, New York, USA. Illus:227-240.

Link T, Backstrom M, Graham R, Essers R, Zorner K, Gatgens J, Burchell J, Taylor-Papadimitriou J, Hansson GC, Noll T. 2004. Bioprocess development for the production of a recombinant MUC1 fusion protein expressed by CHO-K1 cells in protein-free medium. Journal of Biotechnology 110(1):51-62.

Lopez JM, Marques F. 2002. Modulated Taylor-Couette flow: Onset of spiral modes. Theoretical and Computational Fluid Dynamics 16(1):59-69.

Low D, O'Leary R, Pujar NS. 2007. Future of antibody purification. Journal of Chromatography B-Analytical Technologies in the Biomedical and Life Sciences 848(1):48-63.

Ludwig A, Kretzmer G, Schugerl K. 1992. Determination of a Critical Shear-Stress Level Applied to Adherent Mammalian-Cells. Enzyme and Microbial Technology 14(3):209-213.

Lueptow RM, Docter A, Min KY. 1992. Stability of Axial-Flow in an Annulus with a Rotating Inner Cylinder. Physics of Fluids a-Fluid Dynamics 4(11):2446-2455.

Luo J, Yang ST. 2004. Effects of three-dimensional culturing in a fibrous matrix on cell cycle, apoptosis, and MAb production by hybridoma cells. Biotechnology Progress 20(1):306-315.

Lutkemeyer D, Ameskamp N, Tebbe H, Wittler J, Lehmann J. 1999. Estimation of cell damage in bench- and pilot-scale affinity expanded-bed chromatography for the purification of monoclonal antibodies. Biotechnology and Bioengineering 65(1):114-119.

Luttman R, Florek P, Preil W. 1994. Silicone-Tubing Aerated Bioreactors for Somatic Embryo Production. Plant Cell Tissue and Organ Culture 39(2):157-170.

Ma NN, Chalmers JJ, Aunins JG, Zhou WC, Xie LZ. 2004. Quantitative studies of cell-bubble interactions and cell damage at different pluronic F-68 and cell concentrations. Biotechnology Progress 20(4):1183-1191.

Ma NN, Koelling KW, Chalmers JJ. 2002. Fabrication and use of a transient contractional flow device to quantify the sensitivity of mammalian and insect cells to hydrodynamic forces. Biotechnology and Bioengineering 80(4):428-437.

Manna L. 1997. Comparison between physical and chemical methods for the measurement of mixing times. Chemical Engineering Journal 67(3):167-173.

Marks DM. 2003. Equipment design considerations for large scale cell culture. Cytotechnology 42(1):21-33.

Martin SJ, Reutelingsperger CPM, McGahon AJ, Rader JA, Vanschie R, Laface DM, Green DR. 1995. Early Redistribution of Plasma-Membrane Phosphatidylserine Is a General Feature of Apoptosis Regardless of the Initiating Stimulus - Inhibition by Overexpression of Bcl-2 and Abl. Journal of Experimental Medicine 182(5):1545-1556.

Mazur X, Fussenegger M, Renner WA, Bailey JE. 1998. Higher productivity of growth-arrested Chinese hamster ovary cells expressing the cyclin-dependent kinase inhibitor p27. Biotechnology Progress 14(5):705-713.

McQueen A, Bailey JE. 1989. Influence of Serum Level, Cell-Line, Flow Type and Viscosity on Flow-Induced Lysis of Suspended Mammalian-Cells. Biotechnology Letters 11(8):531-536.

McQueen A, Meilhoc E, Bailey JE. 1987. Flow Effects on the Viability and Lysis of Suspended Mammalian-Cells. Biotechnology Letters 9(12):831-836.

Mercille S, Massie B. 1994a. Induction of Apoptosis in Nutrient-Deprived Cultures of Hybridoma and Myeloma Cells. Biotechnology and Bioengineering 44(9):1140-1154.

Mercille S, Massie B. 1994b. Induction of Apoptosis in Oxygen-Deprived Cultures of Hybridoma Cells. Cytotechnology 15(1-3):117-128.

Messi F. 2008. Personal Communication. Zurich, Switzerland.

Michaels JD, Mallik AK, Papoutsakis ET. 1996. Sparging and agitation-induced injury of cultured animal cells: Do cell-to-bubble interactions in the bulk liquid injure cells? Biotechnology and Bioengineering 51(4):399-409.

Michaels JD, Nowak JE, Mallik AK, Koczo K, Wasan DT, Papoutsakis ET. 1995. Interfacial Properties of Cell-Culture Media with Cell-Protecting Additives. Biotechnology and Bioengineering 47(4):420-430.

Michaels JD, Papoutsakis ET. 1991. Polyvinyl-Alcohol and Polyethylene-Glycol as Protectants against Fluid-Mechanical Injury of Freely-Suspended Animal-Cells (Crl-8018). Journal of Biotechnology 19(2-3):241-257.

Miller WM, Wilke CR, Blanch HW. 1987. Effects of Dissolved-Oxygen Concentration on Hybridoma Growth and Metabolism in Continuous Culture. Journal of Cellular Physiology 132(3):524-530.

Millward HR, Bellhouse BJ, Sobey IJ. 1996. The vortex wave membrane bioreactor: Hydrodynamics and mass transfer. Chemical Engineering Journal and the Biochemical Engineering Journal 62(3):175-181.

Mizrahi A. 1975. Pluronic Polyols in Human Lymphocyte Cell Line Cultures. Journal of Clinical Microbiology 2(1):11-13.

Mollet M, Godoy-Silva R, Berdugo C, Chalmers JJ. 2007. Acute hydrodynamic forces and apoptosis: A complex question. Biotechnology and Bioengineering 98:772-788.

Morbidelli M, Soos M, Wu H; Eth Zuerich, assignee. 2005. Reactor and/or mixing vessel for processes e.g. biochemical process comprises cylindrical body with jacket having of irregular cross-section concentrically mounted in cylindrical body having circular cross-section to form non-constant gap patent EP1504808-A1.

Moreira JL, Cruz PE, Santana PC, Feliciano AS, Lehmann J, Carrondo MJT. 1995. Influence of Power Input and Aeration Method on Mass-Transfer in a Laboratory-

Animal Cell-Culture Vessel. Journal of Chemical Technology and Biotechnology 62(2):118-131.

Motobu M, Wang PC, Matsumura M. 1998. Effect of shear stress on recombinant Chinese hamster ovary cells. Journal of Fermentation and Bioengineering 85(2):190-195.

Murhammer DW, Goochee CF. 1988. Scaleup of Insect Cell-Cultures - Protective Effects of Pluronic-F-68. Bio-Technology 6(12):1411-1418.

Murhammer DW, Goochee CF. 1990. Sparged Animal-Cell Bioreactors - Mechanism of Cell-Damage and Pluronic F-68 Protection. Biotechnology Progress 6(5):391-397.

Murhammer DW, Pfalzgraf EC. 1992. Effects of Pluronic F-68 on Oxygen Transport in an Agitated Sparged Bioreactor. Biotechnology Techniques 6(3):199-202.

Nagata S. 1975. Mixing: Principles and Applications. New York: Wiley.

Nehring D, Czermak P, Vorlop J, Lubben H. 2004. Experimental study of a ceramic microsparging aeration system in a pilot-scale animal cell culture. Biotechnology Progress 20(6):1710-1717.

Nienow AW. 1997. On impeller circulation and mixing effectiveness in the turbulent flow regime. Chemical Engineering Science 52(15):2557-2565.

Nienow AW. 2006. Reactor engineering in large scale animal cell culture. Cytotechnology 50(1-3):9-33.

Nienow AW, Langheinrich C, Stevenson NC, Emery AN, Clayton TM, Slater NKH. 1996. Homogenisation and oxygen transfer rates in large agitated and sparged animal cell bioreactors: Some implications for growth and production. Cytotechnology 22(1-3):87-94.

O'Brien IEW, Reutelingsperger CPM, Holdaway KM. 1997. Annexin-V and TUNEL use in monitoring the progression of apoptosis in plants. Cytometry 29(1):28-33.

Ozturk SS, Palsson BO. 1990. Effects of Dissolved-Oxygen on Hybridoma Cell-Growth, Metabolism, and Antibody-Production Kinetics in Continuous Culture. Biotechnology Progress 6(6):437-446.

Palomares LA, Gonzalez M, Ramirez OT. 2000. Evidence of Pluronic F-68 direct interaction with insect cells: impact on shear protection, recombinant protein, and baculovirus production. Enzyme and Microbial Technology 26(5-6):324-331.

Patwardhan AW, Joshi JB. 1999. Relation between flow pattern and blending in stirred tanks. Industrial & Engineering Chemistry Research 38(8):3131-3143.

Pavlou AK, Reichert JM. 2004. Recombinant protein therapeutics - success rates, market trends and values to 2010. Nature Biotechnology 22(12):1513-1519.

Perry RH. 1997. Perry's Chemical Engineers' Handbook 7th Ed.

Persson B, Emborg C. 1992. A Comparison of 3 Different Mammalian-Cell Bioreactors for the Production of Monoclonal-Antibodies. Bioprocess Engineering 8(3-4):157-163.

Petersen JF, McIntire LV, Papoutsakis ET. 1988. Shear Sensitivity of Cultured Hybridoma Cells (Crl-8018) Depends on Mode of Growth, Culture Age and Metabolite Concentration. Journal of Biotechnology 7(3):229-246.

Petit PX, Lecoeur H, Zorn E, Dauguet C, Mignotte B, Gougeon ML. 1995. Alterations in Mitochondrial Structure and Function Are Early Events of Dexamethasone-Induced Thymocyte Apoptosis. Journal of Cell Biology 130(1):157-167.

Poot M, Pierce RH. 1999. Detection of changes in mitochondrial function during apoptosis by simultaneous staining with multiple fluorescent dyes and correlated multiparameter flow cytometry. Cytometry 35(4):311-317.

Pope SB. 2000. Turbulent flows. Cambridge: Cambridge University Press.

Portier BP, Ferrari DC, Taglialatela G. 2006. Rapid assay for quantitative measurement of apoptosis in cultured cells and brain tissue. Journal of Neuroscience Methods 155(1):134-142.

Press WH, Teukolsky SA, Vetterling WT, Flannery BP. 1992. Numerical Recipes in C: The Art of Scientific Computing (2nd edition). Cambridge: Cambridge University Press.

Qi HN, Goudar CT, Michaels JD, Henzler HJ, Jovanovic GN, Konstantinov KB. 2003. Experimental and theoretical analysis of tubular membrane aeration for mammalian cell bioreactors. Biotechnology Progress 19(4):1183-1189.

Racher AJ, Looby D, Griffiths JB. 1990. Use of Lactate-Dehydrogenase Release to Assess Changes in Culture Viability. Cytotechnology 3(3):301-307.

Ramirez OT, Mutharasan R. 1990. The Role of the Plasma-Membrane Fluidity on the Shear Sensitivity of Hybridomas Grown under Hydrodynamic Stress. Biotechnology and Bioengineering 36(9):911-920.

Rasola A, Geuna M. 2001. A flow cytometry assay simultaneously detects independent apoptotic parameters. Cytometry 45(2):151-157.

Rathmell JC, Thompson CB. 1999. The central effectors of cell death in the immune system. Annual Review of Immunology 17:781-828.

Resende MM, Tardioli PW, Fernandez VM, Ferreira ALO, Giordano RLC, Giordano RC. 2001. Distribution of suspended particles in a Taylor-Poiseuille vortex flow reactor. Chemical Engineering Science 56(3):755-761.

Reuveny S, Velez D, Macmillan JD, Miller L. 1986. Factors Affecting Cell-Growth and Monoclonal-Antibody Production in Stirred Reactors. Journal of Immunological Methods 86(1):53-59.

Rheometric Scientific Inc. 1998. Advanced Rheometric Expansion System (ARES) - Instrument Manual. Piscataway, NJ, USA.

Ross LF, Chapital DC. 1987. Simultaneous Determination of Carbohydrates and Products of Carbohydrate-Metabolism in Fermentation Mixtures by Hplc. Journal of Chromatographic Science 25(3):112-117.

Saarinen MA, Murhammer DW. 2000. Culture in the rotating-wall vessel affects recombinant protein production capability of two insect cell lines in different manners. In Vitro Cellular & Developmental Biology-Animal 36(6):362-366.

Sanfeliu A, Stephanopoulos G. 1999. Effect of glutamine limitation on the death of attached Chinese hamster ovary cells. Biotechnology and Bioengineering 64(1):46-53.

Saraste A, Pulkki K. 2000. Morphologic and biochemical hallmarks of apoptosis. Cardiovascular Research 45(3):528-537.

Schlatter S, Rimann M, Kelm J, Fussenegger M. 2002. SAMY, a novel mammalian reporter gene derived from Bacillus stearothermophilus alpha-amylase. Gene 282(1-2):19-31.

Schmid I, Uittenbogaart CH, Giorgi JV. 1994. Sensitive Method for Measuring Apoptosis and Cell-Surface Phenotype in Human Thymocytes by Flow-Cytometry. Cytometry 15(1):12-20.

Schneider M, Reymond F, Marison IW, Vonstockar U. 1995. Bubble-Free Oxygenation by Means of Hydrophobic Porous Membranes. Enzyme and Microbial Technology 17(9):839-847.

Schwarz R, Lewis M, Cross J, Bowie W, Wolf D, Sams C, Cohen G. 1988. Principles and Operation of a Zero Headspace Rotating Wall Cell Culture Apparatus. FASEB Journal 2(4):ABSTRACT 3703.

Senger RS, Karim MN. 2003. Effect of shear stress on intrinsic CHO culture state and glycosylation of recombinant tissue-type plasminogen activator protein. Biotechnology Progress 19(4):1199-1209.

Sengupta TK, Kabir MF, Ray AK. 2001. A Taylor vortex photocatalytic reactor for water purification Industrial Engineering and Chemical Research 40:5268-5281.

Sgonc R, Gruber J. 1998. Apoptosis detection: An overview. Experimental Gerontology 33(6):525-533.

Shive MS, Salloum ML, Anderson JM. 2000. Shear stress-induced apoptosis of adherent neutrophils: A mechanism for persistence of cardiovascular device infections. Proceedings of the National Academy of Sciences of the United States of America 97(12):6710-6715.

Singh RP, Alrubeai M, Gregory CD, Emery AN. 1994. Cell-Death in Bioreactors - a Role for Apoptosis. Biotechnology and Bioengineering 44(6):720-726.

Singh V. 1999. Disposable bioreactor for cell culture using wave-induced agitation. Cytotechnology 30(1-3):149-158.

Smith CG, Greenfield PF, Randerson DH. 1987. A technique for determining the shear sensitivity of mammalian cells in suspension culture Biotechnology Techniques 1(1):39-44.

Snyder HA. 1968. Experiments on Rotating Flows between Noncircular Cylinders. Physics of Fluids 11(8):1606-1611.

Sonntag RC, Russel WB. 1987. Structure and breakup of flocs subjected to fluid stresses III. Converging flow. Journal of Colloid Interface Science 115:390-395.

Soos M, Wu H, Morbidelli M. 2007. Taylor-Couette unit with a lobed inner cylinder cross section. Aiche Journal 53(5):1109-1120.

Stone HA, Bentley BJ, Leal L. 1986. An experimental study of transient effects in the breakup of viscous drops. Journal of Fluid Mechanics 173:131-158.

Sun X, Linden JC. 1999. Shear stress effects on plant cell suspension cultures in a rotating wall vessel bioreactor. Journal of Industrial Microbiology & Biotechnology 22(1):44-47.

Taylor GI. 1923. Stability of a viscous liquid contained between two-rotating cylinders. Philosophical Transactions of the Royal Society of London Philosophical Transactions of the Royal Society of London 223:289-343.

Tennakoon SGK, Andereck CD. 1993. Time-Dependent Patterns in Counterrotating Eccentric Cylinders. Physical Review Letters 71(19):3111-3114.

Tramper J, Williams JB, Joustra D, Vlak JM. 1986. Shear Sensitivity of Insect Cells in Suspension. Enzyme and Microbial Technology 8(1):33-36.

Trump BF, Berezesky IK. 1995. Calcium-Mediated Cell Injury and Cell-Death. Faseb Journal 9(2):219-228.

Van't Riet K. 1979. Review of measuring methods and results in non-viscous gas-liquid mass transfer in stirred vessels. Ind. Eng. Chem. Process Des. Dev. 18:357–364.

van Engeland M, Nieland LJW, Ramaekers FCS, Schutte B, Reutelingsperger CPM. 1998. Annexin V-affinity assay: A review on an apoptosis detection system based on phosphatidylserine exposure. Cytometry 31(1):1-9.

Vedula P, Yeung PK, Fox RO. 2001. Dynamics of scalar dissipation in isotropic turbulence: a numerical and modelling study. Journal of Fluid Mechanics 433:29-60.

Vermes I, Haanen C, Steffensnakken H, Reutelingsperger C. 1995. A Novel Assay for Apoptosis - Flow Cytometric Detection of Phosphatidylserine Expression on Early Apoptotic Cells Using Fluorescein-Labeled Annexin-V. Journal of Immunological Methods 184(1):39-51.

Vohr JH. 1968. An Experimental Study of Taylor Vortices and Turbulence in Flow between Eccentric Rotating Cylinders. Journal of Lubrication Technology 90(1):285-296.

Walsh G. 2006. Biopharmaceutical benchmarks 2006. Nature Biotechnology 24(7):769-776.

Warnock JN, Al-Rubeai M. 2006. Bioreactor systems for the production of biopharmaceuticals from ani mal cells. Biotechnology and Applied Biochemistry 45:1-12.

Weigang F, Reiter M, Jungbauer A, Katinger H. 1989. High-Performance Liquid-Chromatographic Determination of Metabolic Products for Fermentation Control of Mammalian-Cell Culture - Analysis of Carbohydrates, Organic-Acids and

Orthophosphate Using Refractive-Index and Ultraviolet Detectors. Journal of Chromatography-Biomedical Applications 497:59-68.

Wu H, Patterson GK. 1989. Laser-Doppler Measurements of Turbulent-Flow Parameters in a Stirred Mixer. Chemical Engineering Science 44(10):2207-2221.

Wu SC. 1999. Influence of hydrodynamic shear stress on microcarrier-attached cell growth: Cell line dependency and surfactant protection. Bioprocess Engineering 21(3):201-206.

Wurm FM. 2004. Production of recombinant protein therapeutics in cultivated mammalian cells. Nature Biotechnology 22(11):1393-1398.

Wyllie AH. 1980. Glucocorticoid-Induced Thymocyte Apoptosis Is Associated with Endogenous Endonuclease Activation. Nature 284(5756):555-556.

Youn BS, Sen A, Behie LA, Girgis-Gabardo A, Hassell JA. 2006. Scale-up of breast cancer stem cell aggregate cultures to suspension bioreactors. Biotechnology Progress 22(3):801-810.

Zhang G, Yan G, Gurtu V, Spencer C, Kain SR. 1998. Caspase inhibition prevents staurosporine-induced apoptosis in CHO-K1 cells. Apoptosis 3(1):27-33.

Zhang Z, Alrubeai M, Thomas CR. 1993. Estimation of Disruption of Animal-Cells by Turbulent Capillary-Flow. Biotechnology and Bioengineering 42(8):987-993.

Zhou GW, Kresta SM. 1998. Correlation of mean drop size and minimum drop size with the turbulence energy dissipation and the flow in an agitated tank. Chemical Engineering Science 53(11):2063-2079.

Die VDM Verlagsservicegesellschaft sucht für wissenschaftliche Verlage abgeschlossene und herausragende

Dissertationen, Habilitationen, Diplomarbeiten, Master Theses, Magisterarbeiten usw.

für die kostenlose Publikation als Fachbuch.

Sie verfügen über eine Arbeit, die hohen inhaltlichen und formalen Ansprüchen genügt, und haben Interesse an einer honorarvergüteten Publikation?

Dann senden Sie bitte erste Informationen über sich und Ihre Arbeit per Email an *info@vdm-vsg.de*.

Sie erhalten kurzfristig unser Feedback!

VDM Verlagsservicegesellschaft mbH
Dudweiler Landstr. 99   Telefon +49 681 3720 174
D - 66123 Saarbrücken   Fax   +49 681 3720 1749
www.vdm-vsg.de

Die VDM Verlagsservicegesellschaft mbH vertritt

Printed by Books on Demand GmbH, Norderstedt / Germany